DENSITY FUNCTIONAL THEORY

A Practical Introduction

DAVID S. SHOLL
Georgia Institute of Technology

JANICE A. STECKEL
National Energy Technology Laboratory

WILEY

A JOHN WILEY & SONS, INC., PUBLICATION

Prepared in part with support by the National Energy Technology Laboratory

Published by John Wiley & Sons, Inc., Hoboken, New Jersey
Published simultaneously in Canada

For general information on our other products and services or for technical support, please contact our Customer Care Department within the United States at (800) 762-2974, outside the United States at (317) 572-3993 or fax (317) 572-4002.

Wiley also publishes its books in variety of electronic formats. Some content that appears in print may not be available in electronic formats. For more information about Wiley products, visit our web site at www.wiley.com.

Library of Congress Cataloging-in-Publication Data:

Sholl, David S.
  Density functional theory : a practical introduction / David S. Sholl and Jan Steckel.
    p. cm.
  Includes index.
  ISBN 978-0-470-37317-0 (cloth)
  1. Density functionals. 2. Mathematical physics. 3. Quantum chemistry. I. Steckel, Jan.
  II. Title.
    QC20.7.D43S55 2009
    530.14′4—dc22

                                                                                        2008038603

Printed in the United States of America

10 9 8 7

CONTENTS

5   DFT Calculations of Vibrational Frequencies   113

6   Calculating Rates of Chemical Processes Using
Transition State Theory   131

PREFACE

The application of density functional theory (DFT) calculations is rapidly becoming a "standard tool" for diverse materials modeling problems in physics, chemistry, materials science, and multiple branches of engineering. Although a number of highly detailed books and articles on the theoretical foundations of DFT are available, it remains difficult for a newcomer to these methods to rapidly learn the tools that allow him or her to actually perform calculations that are now routine in the fields listed above. This book aims to fill this gap by guiding the reader through the applications of DFT that might be considered the core of continually growing scientific literature based on these methods. Each chapter includes a series of exercises to give readers experience with calculations of their own.

We have aimed to find a balance between brevity and detail that makes it possible for readers to realistically plan to read the entire text. This balance inevitably means certain technical details are explored in a limited way. Our choices have been strongly influenced by our interactions over multiple years with graduate students and postdocs in chemical engineering, physics, chemistry, materials science, and mechanical engineering at Carnegie Mellon University and the Georgia Institute of Technology. A list of Further Reading is provided in each chapter to define appropriate entry points to more detailed treatments of the area. These reading lists should be viewed as identifying highlights in the literature, not as an effort to rigorously cite all relevant work from the thousands of studies that exist on these topics.

One important choice we made to limit the scope of the book was to focus solely on one DFT method suitable for solids and spatially extended materials, namely plane-wave DFT. Although many of the foundations of plane-wave DFT are also relevant to complementary approaches used in the chemistry community for isolated molecules, there are enough differences in the applications of these two groups of methods that including both approaches would only have been possible by significantly expanding the scope of the book. Moreover, several resources already exist that give a practical "hands-on" introduction to computational chemistry calculations for molecules.

Our use of DFT calculations in our own research and our writing of this book has benefited greatly from interactions with numerous colleagues over an extended period. We especially want to acknowledge J. Karl Johnson (University of Pittsburgh), Aravind Asthagiri (University of Florida), Dan Sorescu (National Energy Technology Laboratory), Cathy Stampfl (University of Sydney), John Kitchin (Carnegie Mellon University), and Duane Johnson (University of Illinois). We thank Jeong-Woo Han for his help with a number of the figures. Bill Schneider (University of Notre Dame), Ken Jordan (University of Pittsburgh), and Taku Watanabe (Georgia Institute of Technology) gave detailed and helpful feedback on draft versions. Any errors or inaccuracies in the text are, of course, our responsibility alone.

DSS dedicates this book to his father and father-in-law, whose love of science and curiosity about the world are an inspiration. JAS dedicates this book to her husband, son, and daughter.

DAVID SHOLL

Georgia Institute of Technology,
Atlanta, GA, USA

JAN STECKEL

National Energy Technology Laboratory,
Pittsburgh, PA, USA

1

WHAT IS DENSITY FUNCTIONAL THEORY?

1.1 HOW TO APPROACH THIS BOOK

There are many fields within the physical sciences and engineering where the key to scientific and technological progress is understanding and controlling the properties of matter at the level of individual atoms and molecules. Density functional theory is a phenomenally successful approach to finding solutions to the fundamental equation that describes the quantum behavior of atoms and molecules, the Schrödinger equation, in settings of practical value. This approach has rapidly grown from being a specialized art practiced by a small number of physicists and chemists at the cutting edge of quantum mechanical theory to a tool that is used regularly by large numbers of researchers in chemistry, physics, materials science, chemical engineering, geology, and other disciplines. A search of the *Science Citation Index* for articles published in 1986 with the words "density functional theory" in the title or abstract yields less than 50 entries. Repeating this search for 1996 and 2006 gives more than 1100 and 5600 entries, respectively.

Our aim with this book is to provide just what the title says: an *introduction* to using density functional theory (DFT) calculations in a *practical* context. We do not assume that you have done these calculations before or that you even understand what they are. We do assume that you want to find out what is possible with these methods, either so you can perform calculations

Density Functional Theory: A Practical Introduction. By David S. Sholl and Janice A. Steckel
Copyright © 2009 John Wiley & Sons, Inc.

yourself in a research setting or so you can interact knowledgeably with collaborators who use these methods.

An analogy related to cars may be useful here. Before you learned how to drive, it was presumably clear to you that you can accomplish many useful things with the aid of a car. For you to use a car, it is important to understand the basic concepts that control cars (you need to put fuel in the car regularly, you need to follow basic traffic laws, etc.) and spend time actually driving a car in a variety of road conditions. You do not, however, need to know every detail of how fuel injectors work, how to construct a radiator system that efficiently cools an engine, or any of the other myriad of details that are required if you were going to actually build a car. Many of these details may be important if you plan on undertaking some especially difficult car-related project such as, say, driving yourself across Antarctica, but you can make it across town to a friend's house and back without understanding them.

With this book, we hope you can learn to "drive across town" when doing your own calculations with a DFT package or when interpreting other people's calculations as they relate to physical questions of interest to you. If you are interested in "building a better car" by advancing the cutting edge of method development in this area, then we applaud your enthusiasm. You should continue reading this chapter to find at least one surefire project that could win you a Nobel prize, then delve into the books listed in the Further Reading at the end of the chapter.

At the end of most chapters we have given a series of exercises, most of which involve actually doing calculations using the ideas described in the chapter. Your knowledge and ability will grow most rapidly by doing rather than by simply reading, so we strongly recommend doing as many of the exercises as you can in the time available to you.

1.2 EXAMPLES OF DFT IN ACTION

Before we even define what density functional theory is, it is useful to relate a few vignettes of how it has been used in several scientific fields. We have chosen three examples from three quite different areas of science from the thousands of articles that have been published using these methods. These specific examples have been selected because they show how DFT calculations have been used to make important contributions to a diverse range of compelling scientific questions, generating information that would be essentially impossible to determine through experiments.

1.2.1 Ammonia Synthesis by Heterogeneous Catalysis

Our first example involves an industrial process of immense importance: the catalytic synthesis of ammonia (NH_3). Ammonia is a central component of fertilizers for agriculture, and more than 100 million tons of ammonia are produced commercially each year. By some estimates, more than 1% of all energy used in the world is consumed in the production of ammonia. The core reaction in ammonia production is very simple:

$$N_2 + 3H_2 \longrightarrow 2NH_3.$$

To get this reaction to proceed, the reaction is performed at high temperatures ($>400°C$) and high pressures (>100 atm) in the presence of metals such as iron (Fe) or ruthenium (Ru) that act as catalysts. Although these metal catalysts were identified by Haber and others almost 100 years ago, much is still not known about the mechanisms of the reactions that occur on the surfaces of these catalysts. This incomplete understanding is partly because of the structural complexity of practical catalysts. To make metal catalysts with high surface areas, tiny particles of the active metal are dispersed throughout highly porous materials. This was a widespread application of nanotechnology long before that name was applied to materials to make them sound scientifically exciting! To understand the reactivity of a metal nanoparticle, it is useful to characterize the surface atoms in terms of their local coordination since differences in this coordination can create differences in chemical reactivity; surface atoms can be classified into "types" based on their local coordination. The surfaces of nanoparticles typically include atoms of various types (based on coordination), so the overall surface reactivity is a complicated function of the shape of the nanoparticle and the reactivity of each type of atom.

The discussion above raises a fundamental question: Can a direct connection be made between the shape and size of a metal nanoparticle and its activity as a catalyst for ammonia synthesis? If detailed answers to this question can be found, then they can potentially lead to the synthesis of improved catalysts. One of the most detailed answers to this question to date has come from the DFT calculations of Honkala and co-workers,[1] who studied nanoparticles of Ru. Using DFT calculations, they showed that the net chemical reaction above proceeds via at least 12 distinct steps on a metal catalyst and that the rates of these steps depend strongly on the local coordination of the metal atoms that are involved. One of the most important reactions is the breaking of the N_2 bond on the catalyst surface. On regions of the catalyst surface that were similar to the surfaces of bulk Ru (more specifically, atomically flat regions), a great deal of energy is required for this bond-breaking reaction, implying that the reaction rate is extremely slow. Near Ru atoms that form a common kind of surface step edge on the catalyst, however, a much smaller amount of energy is needed for this reaction. Honkala and co-workers used additional DFT calculations to predict the relative stability of many different local coordinations of surface atoms in Ru nanoparticles in a way that allowed them to predict the detailed shape of the nanoparticles as a function of particle size. This prediction makes a precise connection between the diameter of a Ru nanoparticle and the number of highly desirable reactive sites for breaking N_2 bonds on the nanoparticle. Finally, all of these calculations were used to develop an overall model that describes how the individual reaction rates for the many different kinds of metal atoms on the nanoparticle's surfaces couple together to define the overall reaction rate under realistic reaction conditions. At no stage in this process was any experimental data used to fit or adjust the model, so the final result was a truly predictive description of the reaction rate of a complex catalyst. After all this work was done, Honkala et al. compared their predictions to experimental measurements made with Ru nanoparticle catalysts under reaction conditions similar to industrial conditions. Their predictions were in stunning quantitative agreement with the experimental outcome.

1.2.2   Embrittlement of Metals by Trace Impurities

It is highly likely that as you read these words you are within 1 m of a large number of copper wires since copper is the dominant metal used for carrying electricity between components of electronic devices of all kinds. Aside from its low cost, one of the attractions of copper in practical applications is that it is a soft, ductile metal. Common pieces of copper (and other metals) are almost invariably polycrystalline, meaning that they are made up of many tiny domains called grains that are each well-oriented single crystals. Two neighboring grains have the same crystal structure and symmetry, but their orientation in space is not identical. As a result, the places where grains touch have a considerably more complicated structure than the crystal structure of the pure metal. These regions, which are present in all polycrystalline materials, are called grain boundaries.

It has been known for over 100 years that adding tiny amounts of certain impurities to copper can change the metal from being ductile to a material that will fracture in a brittle way (i.e., without plastic deformation before the fracture). This occurs, for example, when bismuth (Bi) is present in copper (Cu) at levels below 100 ppm. Similar effects have been observed with lead (Pb) or mercury (Hg) impurities. But how does this happen? Qualitatively, when the impurities cause brittle fracture, the fracture tends to occur at grain boundaries, so something about the impurities changes the properties of grain boundaries in a dramatic way. That this can happen at very low concentrations of Bi is not completely implausible because Bi is almost completely insoluble in bulk Cu. This means that it is very favorable for Bi atoms to segregate to grain boundaries rather than to exist inside grains, meaning that the local concentration of Bi at grain boundaries can be much higher than the net concentration in the material as a whole.

Can the changes in copper caused by Bi be explained in a detailed way? As you might expect for an interesting phenomena that has been observed over many years, several alternative explanations have been suggested. One class of explanations assigns the behavior to electronic effects. For example, a Bi atom might cause bonds between nearby Cu atoms to be stiffer than they are in pure Cu, reducing the ability of the Cu lattice to deform smoothly. A second type of electronic effect is that having an impurity atom next to a grain boundary could weaken the bonds that exist across a boundary by changing the electronic structure of the atoms, which would make fracture at the boundary more likely. A third explanation assigns the blame to size effects, noting that Bi atoms are much larger than Cu atoms. If a Bi atom is present at a grain boundary, then it might physically separate Cu atoms on the other side of the boundary from their natural spacing. This stretching of bond distances would weaken the bonds between atoms and make fracture of the grain boundary more likely. Both the second and third explanations involve weakening of bonds near grain boundaries, but they propose different root causes for this behavior. Distinguishing between these proposed mechanisms would be very difficult using direct experiments.

Recently, Schweinfest, Paxton, and Finnis used DFT calculations to offer a definitive description of how Bi embrittles copper; the title of their study gives away the conclusion.[2] They first used DFT to predict stress–strain relationships for pure Cu and Cu containing Bi atoms as impurities. If the bond stiffness argument outlined above was correct, the elastic moduli of the metal should be increased by adding Bi. In fact, the calculations give the opposite result, immediately showing the bond-stiffening explanation to be incorrect. In a separate and much more challenging series of calculations, they explicitly calculated the cohesion energy of a particular grain boundary that is known experimentally to be embrittled by Bi. In qualitative consistency with experimental observations, the calculations predicted that the cohesive energy of the grain boundary is greatly reduced by the presence of Bi. Crucially, the DFT results allow the electronic structure of the grain boundary atoms to be examined directly. The result is that the grain boundary electronic effect outlined above was found to not be the cause of embrittlement. Instead, the large change in the properties of the grain boundary could be understood almost entirely in terms of the excess volume introduced by the Bi atoms, that is, by a size effect. This reasoning suggests that Cu should be embrittled by any impurity that has a much larger atomic size than Cu and that strongly segregates to grain boundaries. This description in fact correctly describes the properties of both Pb and Hg as impurities in Cu, and, as mentioned above, these impurities are known to embrittle Cu.

1.2.3   Materials Properties for Modeling Planetary Formation

To develop detailed models of how planets of various sizes have formed, it is necessary to know (among many other things) what minerals exist inside planets and how effective these minerals are at conducting heat. The extreme conditions that exist inside planets pose some obvious challenges to probing these topics in laboratory experiments. For example, the center of Jupiter has pressures exceeding 40 Mbar and temperatures well above 15,000 K. DFT calculations can play a useful role in probing material properties at these extreme conditions, as shown in the work of Umemoto, Wentzcovitch, and Allen.[3] This work centered on the properties of bulk $MgSiO_3$, a silicate mineral that is important in planet formation. At ambient conditions, $MgSiO_3$ forms a relatively common crystal structure known as a perovskite. Prior to Umemoto et al.'s calculations, it was known that if $MgSiO_3$ was placed under conditions similar to those in the core–mantle boundary of Earth, it transforms into a different crystal structure known as the $CaIrO_3$ structure. (It is conventional to name crystal structures after the first compound discovered with that particular structure, and the naming of this structure is an example of this convention.)

Umemoto et al. wanted to understand what happens to the structure of $MgSiO_3$ at conditions much more extreme than those found in Earth's core–mantle boundary. They used DFT calculations to construct a phase diagram that compared the stability of multiple possible crystal structures of solid $MgSiO_3$. All of these calculations dealt with bulk materials. They also considered the possibility that $MgSiO_3$ might dissociate into other compounds. These calculations predicted that at pressures of ~ 11 Mbar, $MgSiO_3$ dissociates in the following way:

$$MgSiO_3 \text{ [CaIrO}_3 \text{ structure]} \longrightarrow MgO \text{ [CsCl structure]}$$

$$+ SiO_2 \text{ [cotunnite structure]}.$$

In this reaction, the crystal structure of each compound has been noted in the square brackets. An interesting feature of the compounds on the right-hand side is that neither of them is in the crystal structure that is the stable structure at ambient conditions. MgO, for example, prefers the NaCl structure at ambient conditions (i.e., the same crystal structure as everyday table salt). The behavior of SiO_2 is similar but more complicated; this compound goes through several intermediate structures between ambient conditions and the conditions relevant for $MgSiO_3$ dissociation. These transformations in the structures of MgO and SiO_2 allow an important connection to be made between DFT calculations and experiments since these transformations occur at conditions that can be directly probed in laboratory experiments. The transition pressures predicted using DFT and observed experimentally are in good agreement, giving a strong indication of the accuracy of these calculations.

The dissociation reaction predicted by Umemoto et al.'s calculations has important implications for creating good models of planetary formation. At the simplest level, it gives new information about what materials exist inside large planets. The calculations predict, for example, that the center of Uranus or Neptune can contain $MgSiO_3$, but that the cores of Jupiter or Saturn will not. At a more detailed level, the thermodynamic properties of the materials can be used to model phenomena such as convection inside planets. Umemoto et al. speculated that the dissociation reaction above might severely limit convection inside "dense-Saturn," a Saturn-like planet that has been discovered outside the solar system with a mass of $\sim$67 Earth masses.

A legitimate concern about theoretical predictions like the reaction above is that it is difficult to envision how they can be validated against experimental data. Fortunately, DFT calculations can also be used to search for similar types of reactions that occur at pressures that are accessible experimentally. By using this approach, it has been predicted that $NaMgF_3$ goes through a series of transformations similar to $MgSiO_3$; namely, a perovskite to postperovskite transition at some pressure above ambient and then dissociation in NaF and MgF_2 at higher pressures.[4] This dissociation is predicted to occur for pressures around 0.4 Mbar, far lower than the equivalent pressure for $MgSiO_3$. These predictions suggest an avenue for direct experimental tests of the transformation mechanism that DFT calculations have suggested plays a role in planetary formation.

We could fill many more pages with research vignettes showing how DFT calculations have had an impact in many areas of science. Hopefully, these three examples give some flavor of the ways in which DFT calculations can have an impact on scientific understanding. It is useful to think about the common features between these three examples. All of them involve materials in their solid state, although the first example was principally concerned with the interface between a solid and a gas. Each example generated information about a physical problem that is controlled by the properties of materials on atomic length scales that would be (at best) extraordinarily challenging to probe experimentally. In each case, the calculations were used to give information not just about some theoretically ideal state, but instead to understand phenomena at temperatures, pressures, and chemical compositions of direct relevance to physical applications.

1.3 THE SCHRÖDINGER EQUATION

By now we have hopefully convinced you that density functional theory is a useful and interesting topic. But what is it exactly? We begin with the observation that one of the most profound scientific advances of the twentieth century was the development of quantum mechanics and the repeated experimental observations that confirmed that this theory of matter describes, with astonishing accuracy, the universe in which we live.

In this section, we begin a review of some key ideas from quantum mechanics that underlie DFT (and other forms of computational chemistry). Our goal here is not to present a complete derivation of the techniques used in DFT. Instead, our goal is to give a clear, brief, introductory presentation of the most basic equations important for DFT. For the full story, there are a number of excellent texts devoted to quantum mechanics listed in the Further Reading section at the end of the chapter.

Let us imagine a situation where we would like to describe the properties of some well-defined collection of atoms—you could think of an isolated molecule or the atoms defining the crystal of an interesting mineral. One of the fundamental things we would like to know about these atoms is their energy and, more importantly, how their energy changes if we move the atoms around. To define where an atom is, we need to define both where its nucleus is and where the atom's electrons are. A key observation in applying quantum mechanics to atoms is that atomic nuclei are much heavier than individual electrons; each proton or neutron in a nucleus has more than 1800 times the mass of an electron. This means, roughly speaking, that electrons respond much more rapidly to changes in their surroundings than nuclei can. As a result, we can split our physical question into two pieces. First, we solve, for fixed positions of the atomic nuclei, the equations that describe the electron motion. For a given set of electrons moving in the field of a set of nuclei, we find the lowest energy configuration, or *state*, of the electrons. The lowest energy state is known as the *ground state* of the electrons, and the separation of the nuclei and electrons into separate mathematical problems is the *Born–Oppenheimer approximation*. If we have M nuclei at positions $\mathbf{R}_1, \ldots, \mathbf{R}_M$, then we can express the ground-state energy, E, as a function of the positions of these nuclei, $E(\mathbf{R}_1, \ldots, \mathbf{R}_M)$. This function is known as the *adiabatic potential energy surface* of the atoms. Once we are able to calculate this potential energy surface we can tackle the original problem posed above—how does the energy of the material change as we move its atoms around?

One simple form of the Schrödinger equation—more precisely, the time-independent, nonrelativistic Schrödinger equation—you may be familiar with is $H\psi = E\psi$. This equation is in a nice form for putting on a T-shirt or a coffee mug, but to understand it better we need to define the quantities that appear in it. In this equation, H is the Hamiltonian operator and ψ is a set of solutions, or eigenstates, of the Hamiltonian. Each of these solutions,

ψ_n, has an associated eigenvalue, E_n, a real number* that satisfies the eigenvalue equation. The detailed definition of the Hamiltonian depends on the physical system being described by the Schrödinger equation. There are several well-known examples like the particle in a box or a harmonic oscillator where the Hamiltonian has a simple form and the Schrödinger equation can be solved exactly. The situation we are interested in where multiple electrons are interacting with multiple nuclei is more complicated. In this case, a more complete description of the Schrödinger is

$$\left[-\frac{\hbar^2}{2m} \sum_{i=1}^{N} \nabla_i^2 + \sum_{i=1}^{N} V(\mathbf{r}_i) + \sum_{i=1}^{N} \sum_{j<i} U(\mathbf{r}_i, \mathbf{r}_j) \right] \psi = E\psi. \qquad (1.1)$$

Here, m is the electron mass. The three terms in brackets in this equation define, in order, the kinetic energy of each electron, the interaction energy between each electron and the collection of atomic nuclei, and the interaction energy between different electrons. For the Hamiltonian we have chosen, ψ is the electronic wave function, which is a function of each of the spatial coordinates of each of the N electrons, so $\psi = \psi(\mathbf{r}_1, \ldots, \mathbf{r}_N)$, and E is the ground-state energy of the electrons.** The ground-state energy is independent of time, so this is the time-independent Schrödinger equation.†

Although the electron wave function is a function of each of the coordinates of all N electrons, it is possible to approximate ψ as a product of individual electron wave functions, $\psi = \psi_1(\mathbf{r})\psi_2(\mathbf{r}), \ldots, \psi_N(\mathbf{r})$. This expression for the wave function is known as a Hartree product, and there are good motivations for approximating the full wave function into a product of individual one-electron wave functions in this fashion. Notice that N, the number of electrons, is considerably larger than M, the number of nuclei, simply because each atom has one nucleus and lots of electrons. If we were interested in a single molecule of CO_2, the full wave function is a 66-dimensional function (3 dimensions for each of the 22 electrons). If we were interested in a nanocluster of 100 Pt atoms, the full wave function requires more the 23,000 dimensions! These numbers should begin to give you an idea about why solving the Schrödinger equation for practical materials has occupied many brilliant minds for a good fraction of a century.

---

*The value of the functions ψ_n are complex numbers, but the eigenvalues of the Schrödinger equation are real numbers.

**For clarity of presentation, we have neglected electron spin in our description. In a complete presentation, each electron is defined by three spatial variables and its spin.

†The dynamics of electrons are defined by the time-dependent Schrödinger equation, $i\hbar(\partial\psi/\partial t) = H\psi$. The appearance of $i = \sqrt{-1}$ in this equation makes it clear that the wave function is a complex-valued function, not a real-valued function.

The situation looks even worse when we look again at the Hamiltonian, H. The term in the Hamiltonian defining electron–electron interactions is the most critical one from the point of view of solving the equation. The form of this contribution means that the individual electron wave function we defined above, $\psi_i(\mathbf{r})$, cannot be found without simultaneously considering the individual electron wave functions associated with all the other electrons. In other words, the Schrödinger equation is a many-body problem.

Although solving the Schrödinger equation can be viewed as the fundamental problem of quantum mechanics, it is worth realizing that the wave function for any particular set of coordinates cannot be directly observed. The quantity that can (in principle) be measured is the probability that the N electrons are at a particular set of coordinates, $\mathbf{r}_1, \ldots, \mathbf{r}_N$. This probability is equal to $\psi^*(\mathbf{r}_1, \ldots, \mathbf{r}_N)\psi(\mathbf{r}_1, \ldots, \mathbf{r}_N)$, where the asterisk indicates a complex conjugate. A further point to notice is that in experiments we typically do not care which electron in the material is labeled electron 1, electron 2, and so on. Moreover, even if we did care, we cannot easily assign these labels. This means that the quantity of physical interest is really the probability that a set of N electrons in any order have coordinates $\mathbf{r}_1, \ldots, \mathbf{r}_N$. A closely related quantity is the density of electrons at a particular position in space, $n(\mathbf{r})$. This can be written in terms of the individual electron wave functions as

$$n(\mathbf{r}) = 2 \sum_i \psi_i^*(\mathbf{r})\psi_i(\mathbf{r}). \qquad (1.2)$$

Here, the summation goes over all the individual electron wave functions that are occupied by electrons, so the term inside the summation is the probability that an electron in individual wave function $\psi_i(\mathbf{r})$ is located at position $\mathbf{r}$. The factor of 2 appears because electrons have spin and the Pauli exclusion principle states that each individual electron wave function can be occupied by two separate electrons provided they have different spins. This is a purely quantum mechanical effect that has no counterpart in classical physics. The point of this discussion is that the electron density, $n(\mathbf{r})$, which is a function of only three coordinates, contains a great amount of the information that is actually physically observable from the full wave function solution to the Schrödinger equation, which is a function of $3N$ coordinates.

1.4 DENSITY FUNCTIONAL THEORY—FROM WAVE FUNCTIONS TO ELECTRON DENSITY

The entire field of density functional theory rests on two fundamental mathematical theorems proved by Kohn and Hohenberg and the derivation of a set of equations by Kohn and Sham in the mid-1960s. The first theorem, proved by Hohenberg and Kohn, is: *The ground-state energy from Schrödinger's equation is a unique functional of the electron density.*

This theorem states that there exists a one-to-one mapping between the ground-state wave function and the ground-state electron density. To appreciate the importance of this result, you first need to know what a "functional" is. As you might guess from the name, a functional is closely related to the more familiar concept of a function. A function takes a value of a variable or variables and defines a single number from those variables. A simple example of a function dependent on a single variable is $f(x) = x^2 + 1$. A *functional* is similar, but it takes a function and defines a single number from the function. For example,

$$F[f] = \int_{-1}^{1} f(x) \, dx,$$

is a functional of the function $f(x)$. If we evaluate this functional using $f(x) = x^2 + 1$, we get $F[f] = \frac{8}{3}$. So we can restate Hohenberg and Kohn's result by saying that the ground-state energy E can be expressed as $E[n(\mathbf{r})]$, where $n(\mathbf{r})$ is the electron density. This is why this field is known as density functional theory.

Another way to restate Hohenberg and Kohn's result is that the ground-state electron density uniquely determines all properties, including the energy and wave function, of the ground state. Why is this result important? It means that we can think about solving the Schrödinger equation by finding a function of three spatial variables, the electron density, rather than a function of $3N$ variables, the wave function. Here, by "solving the Schrödinger equation" we mean, to say it more precisely, finding the ground-state energy. So for a nanocluster of 100 Pd atoms the theorem reduces the problem from something with more than 23,000 dimensions to a problem with just 3 dimensions.

Unfortunately, although the first Hohenberg–Kohn theorem rigorously proves that a functional of the electron density exists that can be used to solve the Schrödinger equation, the theorem says nothing about what the functional actually is. The second Hohenberg–Kohn theorem defines an important property of the functional: *The electron density that minimizes the energy of the overall functional is the true electron density corresponding to the full solution of the Schrödinger equation.* If the "true" functional form were known, then we could vary the electron density until the energy from the functional is minimized, giving us a prescription for finding the relevant electron density. This variational principle is used in practice with approximate forms of the functional.

A useful way to write down the functional described by the Hohenberg–Kohn theorem is in terms of the single-electron wave functions, $\psi_i(\mathbf{r})$. Remember from Eq. (1.2) that these functions collectively define the electron density, $n(\mathbf{r})$. The energy functional can be written as

$$E[\{\psi_i\}] = E_{\text{known}}[\{\psi_i\}] + E_{\text{XC}}[\{\psi_i\}], \qquad (1.3)$$

where we have split the functional into a collection of terms we can write down in a simple analytical form, $E_{\text{known}}[\{\psi_i\}]$, and everything else, E_{XC}. The "known" terms include four contributions:

$$E_{\text{known}}[\{\psi_i\}] = -\frac{\hbar^2}{m}\sum_i \int \psi_i^* \nabla^2 \psi_i d^3 r + \int V(\mathbf{r})n(\mathbf{r})\, d^3 r$$

$$+\frac{e^2}{2}\int\int \frac{n(\mathbf{r})n(\mathbf{r}')}{|\mathbf{r}-\mathbf{r}'|}\, d^3 r\, d^3 r' + E_{\text{ion}}. \qquad (1.4)$$

The terms on the right are, in order, the electron kinetic energies, the Coulomb interactions between the electrons and the nuclei, the Coulomb interactions between pairs of electrons, and the Coulomb interactions between pairs of nuclei. The other term in the complete energy functional, $E_{\text{XC}}[\{\psi_i\}]$, is the exchange–correlation functional, and it is defined to include all the quantum mechanical effects that are not included in the "known" terms.

Let us imagine for now that we can express the as-yet-undefined exchange–correlation energy functional in some useful way. What is involved in finding minimum energy solutions of the total energy functional? Nothing we have presented so far really guarantees that this task is any easier than the formidable task of fully solving the Schrödinger equation for the wave function. This difficulty was solved by Kohn and Sham, who showed that the task of finding the right electron density can be expressed in a way that involves solving a set of equations in which each equation only involves a single electron.

The Kohn–Sham equations have the form

$$\left[-\frac{\hbar^2}{2m}\nabla^2 + V(\mathbf{r}) + V_H(\mathbf{r}) + V_{\text{XC}}(\mathbf{r}) \right]\psi_i(\mathbf{r}) = \varepsilon_i\psi_i(\mathbf{r}). \qquad (1.5)$$

These equations are superficially similar to Eq. (1.1). The main difference is that the Kohn–Sham equations are missing the summations that appear inside the full Schrödinger equation [Eq. (1.1)]. This is because the solution of the Kohn–Sham equations are single-electron wave functions that depend on only three spatial variables, $\psi_i(\mathbf{r})$. On the left-hand side of the Kohn–Sham equations there are three potentials, V, V_H, and V_{XC}. The first of these also appeared in the full Schrödinger equation (Eq. (1.1)) and in the "known" part of the total energy functional given above (Eq. (1.4)). This potential defines the interaction between an electron and the collection of atomic nuclei. The second is called the Hartree potential and is defined by

$$V_H(\mathbf{r}) = e^2 \int \frac{n(\mathbf{r}')}{|\mathbf{r} - \mathbf{r}'|} d^3 r'.$$ (1.6)

This potential describes the Coulomb repulsion between the electron being considered in one of the Kohn–Sham equations and the total electron density defined by all electrons in the problem. The Hartree potential includes a so-called self-interaction contribution because the electron we are describing in the Kohn–Sham equation is also part of the total electron density, so part of V_H involves a Coulomb interaction between the electron and itself. The self-interaction is unphysical, and the correction for it is one of several effects that are lumped together into the final potential in the Kohn–Sham equations, V_{XC}, which defines exchange and correlation contributions to the single-electron equations. V_{XC} can formally be defined as a "functional derivative" of the exchange–correlation energy:

$$V_{XC}(\mathbf{r}) = \frac{\delta E_{XC}(\mathbf{r})}{\delta n(\mathbf{r})}.$$ (1.7)

The strict mathematical definition of a functional derivative is slightly more subtle than the more familiar definition of a function's derivative, but conceptually you can think of this just as a regular derivative. The functional derivative is written using δ rather than d to emphasize that it not quite identical to a normal derivative.

If you have a vague sense that there is something circular about our discussion of the Kohn–Sham equations you are exactly right. To solve the Kohn–Sham equations, we need to define the Hartree potential, and to define the Hartree potential we need to know the electron density. But to find the electron density, we must know the single-electron wave functions, and to know these wave functions we must solve the Kohn–Sham equations. To break this circle, the problem is usually treated in an iterative way as outlined in the following algorithm:

1. Define an initial, trial electron density, $n(\mathbf{r})$.
2. Solve the Kohn–Sham equations defined using the trial electron density to find the single-particle wave functions, $\psi_i(\mathbf{r})$.
3. Calculate the electron density defined by the Kohn–Sham single-particle wave functions from step 2, $n_{KS}(\mathbf{r}) = 2 \sum_i \psi_i^*(\mathbf{r}) \psi_i(\mathbf{r})$.

4. Compare the calculated electron density, $n_{KS}(\mathbf{r})$, with the electron density used in solving the Kohn–Sham equations, $n(\mathbf{r})$. If the two densities are the same, then this is the ground-state electron density, and it can be used to compute the total energy. If the two densities are different, then the trial electron density must be updated in some way. Once this is done, the process begins again from step 2.

We have skipped over a whole series of important details in this process (How close do the two electron densities have to be before we consider them to be the same? What is a good way to update the trial electron density? How should we define the initial density?), but you should be able to see how this iterative method can lead to a solution of the Kohn–Sham equations that is *self-consistent*.

1.5   EXCHANGE–CORRELATION FUNCTIONAL

Let us briefly review what we have seen so far. We would like to find the ground-state energy of the Schrödinger equation, but this is extremely difficult because this is a many-body problem. The beautiful results of Kohn, Hohenberg, and Sham showed us that the ground state we seek can be found by minimizing the energy of an energy functional, and that this can be achieved by finding a self-consistent solution to a set of single-particle equations. There is just one critical complication in this otherwise beautiful formulation: to solve the Kohn–Sham equations we must specify the exchange–correlation function, $E_{XC}[\{\psi_i\}]$. As you might gather from Eqs. (1.3) and (1.4), defining $E_{XC}[\{\psi_i\}]$ is very difficult. After all, the whole point of Eq. (1.4) is that we have already explicitly written down all the "easy" parts.

In fact, the true form of the exchange–correlation functional whose existence is guaranteed by the Hohenberg–Kohn theorem is simply not known. Fortunately, there is one case where this functional can be derived exactly: the uniform electron gas. In this situation, the electron density is constant at all points in space; that is, $n(\mathbf{r}) = $ constant. This situation may appear to be of limited value in any real material since it is variations in electron density that define chemical bonds and generally make materials interesting. But the uniform electron gas provides a practical way to actually use the Kohn–Sham equations. To do this, we set the exchange–correlation potential at each position to be the known exchange–correlation potential from the uniform electron gas at the electron density observed at that position:

$$V_{XC}(\mathbf{r}) = V_{XC}^{\text{electron gas}}[n(\mathbf{r})]. \tag{1.8}$$

This approximation uses only the local density to define the approximate exchange–correlation functional, so it is called the *local density approximation* (LDA). The LDA gives us a way to completely define the Kohn–Sham equations, but it is crucial to remember that the results from these equations do not exactly solve the true Schrödinger equation because we are not using the true exchange–correlation functional.

It should not surprise you that the LDA is not the only functional that has been tried within DFT calculations. The development of functionals that more faithfully represent nature remains one of the most important areas of active research in the quantum chemistry community. We promised at the beginning of the chapter to pose a problem that could win you the Nobel prize. Here it is: Develop a functional that accurately represents nature's exact functional and implement it in a mathematical form that can be efficiently solved for large numbers of atoms. (This advice is a little like the Hohenberg–Kohn theorem—it tells you that something exists without providing any clues how to find it.)

Even though you could become a household name (at least in scientific circles) by solving this problem rigorously, there are a number of approximate functionals that have been found to give good results in a large variety of physical problems and that have been widely adopted. The primary aim of this book is to help you understand how to do calculations with these existing functionals. The best known class of functional after the LDA uses information about the local electron density and the local gradient in the electron density; this approach defines a *generalized gradient approximation* (GGA). It is tempting to think that because the GGA includes more physical information than the LDA it must be more accurate. Unfortunately, this is not always correct.

Because there are many ways in which information from the gradient of the electron density can be included in a GGA functional, there are a large number of distinct GGA functionals. Two of the most widely used functionals in calculations involving solids are the Perdew–Wang functional (PW91) and the Perdew–Burke–Ernzerhof functional (PBE). Each of these functionals are GGA functionals, and dozens of other GGA functionals have been developed and used, particularly for calculations with isolated molecules. Because different functionals will give somewhat different results for any particular configuration of atoms, it is necessary to specify what functional was used in any particular calculation rather than simple referring to "a DFT calculation."

Our description of GGA functionals as including information from the electron density and the gradient of this density suggests that more sophisticated functionals can be constructed that use other pieces of physical information. In fact, a hierarchy of functionals can be constructed that gradually include more and more detailed physical information. More information about this hierarchy of functionals is given in Section 10.2.

1.6   THE QUANTUM CHEMISTRY TOURIST

As you read about the approaches aside from DFT that exist for finding numerical solutions of the Schrödinger equation, it is likely that you will rapidly encounter a bewildering array of acronyms. This experience could be a little bit like visiting a sophisticated city in an unfamiliar country. You may recognize that this new city is beautiful, and you definitely wish to appreciate its merits, but you are not planning to live there permanently. You could spend years in advance of your trip studying the language, history, culture, and geography of the country before your visit, but most likely for a brief visit you are more interested in talking with some friends who have already visited there, reading a few travel guides, browsing a phrase book, and perhaps trying to identify a few good local restaurants. This section aims to present an overview of quantum chemical methods on the level of a phrase book or travel guide.

1.6.1   Localized and Spatially Extended Functions

One useful way to classify quantum chemistry calculations is according to the types of functions they use to represent their solutions. Broadly speaking, these methods use either spatially localized functions or spatially extended functions. As an example of a spatially localized function, Fig. 1.1 shows the function

$$f(x) = f_1(x) + f_2(x) + f_3(x), \tag{1.9}$$

where $f_1(x) = \exp(-x^2),$
$f_2(x) = x^2 \exp(-x^2/2),$
$f_3(x) = \frac{1}{10} x^2 (1 - x)^2 \exp(-x^2/4).$

Figure 1.1 also shows $f_1, f_2,$ and f_3. All of these functions rapidly approach zero for large values of $|x|$. Functions like this are entirely appropriate for representing the wave function or electron density of an isolated atom. This example incorporates the idea that we can combine multiple individual functions with different spatial extents, symmetries, and so on to define an overall function. We could include more information in this final function by including more individual functions within its definition. Also, we could build up functions that describe multiple atoms simply by using an appropriate set of localized functions for each individual atom.

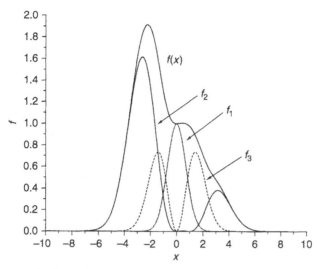

Figure 1.1   Example of spatially localized functions defined in the text.

Spatially localized functions are an extremely useful framework for thinking about the quantum chemistry of isolated molecules because the wave functions of isolated molecules really do decay to zero far away from the molecule. But what if we are interested in a bulk material such as the atoms in solid silicon or the atoms beneath the surface of a metal catalyst? We could still use spatially localized functions to describe each atom and add up these functions to describe the overall material, but this is certainly not the only way forward. A useful alternative is to use periodic functions to describe the wave functions or electron densities. Figure 1.2 shows a simple example of this idea by plotting

$$f(x) = f_1(x) + f_2(x) + f_3(x),$$

where $f_1(x) = \sin^2\left(\dfrac{\pi x}{4}\right),$

$f_2(x) = \dfrac{1}{3}\cos^2\left(\dfrac{\pi x}{2}\right),$

$f_3(x) = \frac{1}{10}\sin^2(\pi x).$

The resulting function is periodic; that is

$$f(x + 4n) = f(x),$$

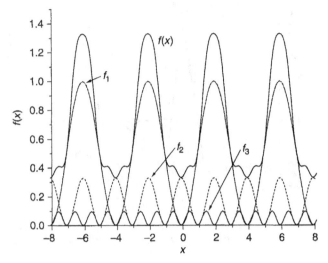

Figure 1.2   Example of spatially periodic functions defined in the text.

for any integer n. This type of function is useful for describing bulk materials since at least for defect-free materials the electron density and wave function really are spatially periodic functions.

Because spatially localized functions are the natural choice for isolated molecules, the quantum chemistry methods developed within the chemistry community are dominated by methods based on these functions. Conversely, because physicists have historically been more interested in bulk materials than in individual molecules, numerical methods for solving the Schrödinger equation developed in the physics community are dominated by spatially periodic functions. You should not view one of these approaches as "right" and the other as "wrong" as they both have advantages and disadvantages.

1.6.2   Wave-Function-Based Methods

A second fundamental classification of quantum chemistry calculations can be made according to the quantity that is being calculated. Our introduction to DFT in the previous sections has emphasized that in DFT the aim is to compute the electron *density*, not the electron *wave function*. There are many methods, however, where the object of the calculation is to compute the full electron wave function. These wave-function-based methods hold a crucial advantage over DFT calculations in that there is a well-defined hierarchy of methods that, given infinite computer time, can converge to the *exact* solution of the Schrödinger equation. We cannot do justice to the breadth of this field in just a few paragraphs, but several excellent introductory texts are available and are listed in the Further Reading section at the end of this chapter. The strong connections between DFT and wave-function-based methods and their importance together within science was recognized in 1998 when the Nobel prize in chemistry was awarded jointly to Walter Kohn for his work developing the foundations of DFT and John Pople for his groundbreaking work on developing a quantum chemistry computer code for calculating the electronic structure of atoms and molecules. It is interesting to note that this was the first time that a Nobel prize in chemistry or physics was awarded for the development of a numerical method (or more precisely, a class of numerical methods) rather than a distinct scientific discovery. Kohn's Nobel lecture gives a very readable description of the advantages and disadvantages of wave-function-based and DFT calculations.[5]

Before giving a brief discussion of wave-function-based methods, we must first describe the common ways in which the wave function is described. We mentioned earlier that the wave function of an N-particle system is an N-dimensional function. But what, exactly, *is* a wave function? Because we want our wave functions to provide a quantum mechanical description of a system of N electrons, these wave functions must satisfy several mathematical properties exhibited by real electrons. For example, the Pauli exclusion principle prohibits two electrons with the same spin from existing at the same physical location simultaneously.[‡] We would, of course, like these properties to also exist in any approximate form of the wave function that we construct.

1.6.3 Hartree–Fock Method

Suppose we would like to approximate the wave function of N electrons. Let us assume for the moment that the electrons have *no effect* on each other. If this is true, the Hamiltonian for the electrons may be written as

$$H = \sum_{i=1}^{N} h_i, \qquad (1.10)$$

where h_i describes the kinetic and potential energy of electron i. The full electronic Hamiltonian we wrote down in Eq. (1.1) takes this form if we simply neglect electron–electron interactions. If we write down the Schrödinger

---

[‡]Spin is a quantum mechanical property that does not appear in classical mechanics. An electron can have one of two distinct spins, spin up or spin down. The full specification of an electron's state must include both its location and its spin. The Pauli exclusion principle only applies to electrons with the same spin state.

equation for just one electron based on this Hamiltonian, the solutions would satisfy

$$h\chi = E\chi. \tag{1.11}$$

The eigenfunctions defined by this equation are called *spin orbitals*. For each single-electron equation there are multiple eigenfunctions, so this defines a set of spin orbitals $\chi_j(\mathbf{x}_i)$ ($j = 1, 2, \ldots$) where $\mathbf{x}_i$ is a vector of coordinates that defines the position of electron i and its spin state (up or down). We will denote the energy of spin orbital $\chi_j(\mathbf{x}_i)$ by E_j. It is useful to label the spin orbitals so that the orbital with $j = 1$ has the lowest energy, the orbital with $j = 2$ has the next highest energy, and so on. When the total Hamiltonian is simply a sum of one-electron operators, h_i, it follows that the eigenfunctions of H are products of the one-electron spin orbitals:

$$\psi(\mathbf{x}_1, \ldots, \mathbf{x}_N) = \chi_{j_1}(\mathbf{x}_1)\chi_{j_2}(\mathbf{x}_2)\cdots\chi_{j_N}(\mathbf{x}_N). \tag{1.12}$$

The energy of this wave function is the sum of the spin orbital energies, $E = E_{j_1} + \cdots + E_{j_N}$. We have already seen a brief glimpse of this approximation to the N-electron wave function, the Hartree product, in Section 1.3.

   Unfortunately, the Hartree product does not satisfy all the important criteria for wave functions. Because electrons are fermions, the wave function must change sign if two electrons change places with each other. This is known as the antisymmetry principle. Exchanging two electrons does not change the sign of the Hartree product, which is a serious drawback. We can obtain a better approximation to the wave function by using a Slater determinant. In a Slater determinant, the N-electron wave function is formed by combining one-electron wave functions in a way that satisfies the antisymmetry principle. This is done by expressing the overall wave function as the determinant of a matrix of single-electron wave functions. It is best to see how this works for the case of two electrons. For two electrons, the Slater determinant is

$$\psi(\mathbf{x}_1, \mathbf{x}_2) = \frac{1}{\sqrt{2}}\det\begin{bmatrix} \chi_j(\mathbf{x}_1) & \chi_j(\mathbf{x}_2) \\ \chi_k(\mathbf{x}_1) & \chi_k(\mathbf{x}_1) \end{bmatrix}$$

$$= \frac{1}{\sqrt{2}}\left[\chi_j(\mathbf{x}_1)\chi_k(\mathbf{x}_1) - \chi_j(\mathbf{x}_2)\chi_k(\mathbf{x}_1)\right]. \tag{1.13}$$

The coefficient of $(1/\sqrt{2})$ is simply a normalization factor. This expression builds in a physical description of electron exchange implicitly; it changes sign if two electrons are exchanged. This expression has other advantages. For example, it does not distinguish between electrons and it disappears if two electrons have the same coordinates or if two of the one-electron wave functions are the same. This means that the Slater determinant satisfies the conditions of the Pauli exclusion principle. The Slater determinant may be generalized to a system of N electrons easily; it is the determinant of an $N \times N$ matrix of single-electron spin orbitals. By using a Slater determinant, we are ensuring that our method for solving the Schrödinger equation will include exchange. Unfortunately, this is not the only kind of electron correlation that we need to describe in order to arrive at good computational accuracy.

The description above may seem a little unhelpful since we know that in any interesting system the electrons interact with one another. The many different wave-function-based approaches to solving the Schrödinger equation differ in how these interactions are approximated. To understand the types of approximations that can be used, it is worth looking at the simplest approach, the Hartree–Fock method, in some detail. There are also many similarities between Hartree–Fock calculations and the DFT calculations we have described in the previous sections, so understanding this method is a useful way to view these ideas from a slightly different perspective.

In a Hartree–Fock (HF) calculation, we fix the positions of the atomic nuclei and aim to determine the wave function of N-interacting electrons. The first part of describing an HF calculation is to define what equations are solved. The Schrödinger equation for each electron is written as

$$\left[-\frac{\hbar^2}{2m} \nabla^2 + V(\mathbf{r}) + V_H(\mathbf{r}) \right] \chi_j(\mathbf{x}) = E_j \chi_j(\mathbf{x}). \tag{1.14}$$

The third term on the left-hand side is the same Hartree potential we saw in Eq. (1.5):

$$V_H(\mathbf{r}) = e^2 \int \frac{n(\mathbf{r}')}{|\mathbf{r} - \mathbf{r}'|} d^3 r'. \tag{1.15}$$

In plain language, this means that a single electron "feels" the effect of other electrons only as an average, rather than feeling the instantaneous repulsive forces generated as electrons become close in space. If you compare Eq. (1.14) with the Kohn–Sham equations, Eq. (1.5), you will notice that the only difference between the two sets of equations is the additional exchange–correlation potential that appears in the Kohn–Sham equations.

To complete our description of the HF method, we have to define how the solutions of the single-electron equations above are expressed and how these solutions are combined to give the N-electron wave function. The HF approach assumes that the complete wave function can be approximated using a single Slater determinant. This means that the N lowest energy spin orbitals of the single-electron equation are found, $\chi_j(\mathbf{x})$ for $j = 1, \ldots, N$, and the total wave function is formed from the Slater determinant of these spin orbitals.

To actually solve the single-electron equation in a practical calculation, we have to define the spin orbitals using a finite amount of information since we cannot describe an arbitrary continuous function on a computer. To do this, we define a finite set of functions that can be added together to approximate the exact spin orbitals. If our finite set of functions is written as $\phi_1(\mathbf{x}), \phi_2(\mathbf{x}), \ldots, \phi_K(\mathbf{x})$, then we can approximate the spin orbitals as

$$\chi_j(\mathbf{x}) = \sum_{i=1}^{K} \alpha_{j,i}\phi_i(\mathbf{x}). \tag{1.16}$$

When using this expression, we only need to find the expansion coefficients, $\alpha_{j,i}$, for $i = 1, \ldots, K$ and $j = 1, \ldots, N$ to fully define all the spin orbitals that are used in the HF method. The set of functions $\phi_1(\mathbf{x}), \phi_2(\mathbf{x}), \ldots, \phi_K(\mathbf{x})$ is called the *basis set* for the calculation. Intuitively, you can guess that using a larger basis set (i.e., increasing K) will increase the accuracy of the calculation but also increase the amount of effort needed to find a solution. Similarly, choosing basis functions that are very similar to the types of spin orbitals that actually appear in real materials will improve the accuracy of an HF calculation. As we hinted at in Section 1.6.1, the characteristics of these functions can differ depending on the type of material that is being considered.

We now have all the pieces in place to perform an HF calculation—a basis set in which the individual spin orbitals are expanded, the equations that the spin orbitals must satisfy, and a prescription for forming the final wave function once the spin orbitals are known. But there is one crucial complication left to deal with; one that also appeared when we discussed the Kohn–Sham equations in Section 1.4. To find the spin orbitals we must solve the single-electron equations. To define the Hartree potential in the single-electron equations, we must know the electron density. But to know the electron density, we must define the electron wave function, which is found using the individual spin orbitals! To break this circle, an HF calculation is an iterative procedure that can be outlined as follows:

1. Make an initial estimate of the spin orbitals $\chi_j(\mathbf{x}) = \sum_{i=1}^{K} \alpha_{j,i}\phi_i(\mathbf{x})$ by specifying the expansion coefficients, $\alpha_{j,i}$.
2. From the current estimate of the spin orbitals, define the electron density, $n(\mathbf{r}')$.
3. Using the electron density from step 2, solve the single-electron equations for the spin orbitals.

4. If the spin orbitals found in step 3 are consistent with orbitals used in step 2, then these are the solutions to the HF problem we set out to calculate. If not, then a new estimate for the spin orbitals must be made and we then return to step 2.

This procedure is extremely similar to the iterative method we outlined in Section 1.4 for solving the Kohn–Sham equations within a DFT calculation. Just as in our discussion in Section 1.4, we have glossed over many details that are of great importance for actually doing an HF calculation. To identify just a few of these details: How do we decide if two sets of spin orbitals are similar enough to be called consistent? How can we update the spin orbitals in step 4 so that the overall calculation will actually converge to a solution? How large should a basis set be? How can we form a useful initial estimate of the spin orbitals? How do we efficiently find the expansion coefficients that define the solutions to the single-electron equations? Delving into the details of these issues would take us well beyond our aim in this section of giving an overview of quantum chemistry methods, but we hope that you can appreciate that reasonable answers to each of these questions can be found that allow HF calculations to be performed for physically interesting materials.

1.6.4 Beyond Hartree–Fock

The Hartree–Fock method provides an exact description of electron exchange. This means that wave functions from HF calculations have exactly the same properties when the coordinates of two or more electrons are exchanged as the true solutions of the full Schrödinger equation. If HF calculations were possible using an infinitely large basis set, the energy of N electrons that would be calculated is known as the *Hartree–Fock limit*. This energy is not the same as the energy for the true electron wave function because the HF method does not correctly describe how electrons influence other electrons. More succinctly, the HF method does not deal with electron correlations.

As we hinted at in the previous sections, writing down the physical laws that govern electron correlation is straightforward, but finding an exact description of electron correlation is intractable for any but the simplest systems. For the purposes of quantum chemistry, the energy due to electron correlation is defined in a specific way: the electron correlation energy is the difference between the Hartree–Fock limit and the true (non-relativistic) ground-state energy. Quantum chemistry approaches that are more sophisticated than the HF method for approximately solving the Schrödinger equation capture some part of the electron correlation energy by improving in some way upon one of the assumptions that were adopted in the Hartree–Fock approach.

How do more advanced quantum chemical approaches improve on the HF method? The approaches vary, but the common goal is to include a description of electron correlation. Electron correlation is often described by "mixing" into the wave function some configurations in which electrons have been excited or promoted from lower energy to higher energy orbitals. One group of methods that does this are the single-determinant methods in which a single Slater determinant is used as the reference wave function and excitations are made from that wave function. Methods based on a single reference determinant are formally known as "post–Hartree–Fock" methods. These methods include configuration interaction (CI), coupled cluster (CC), Møller–Plesset perturbation theory (MP), and the quadratic configuration interaction (QCI) approach. Each of these methods has multiple variants with names that describe salient details of the methods. For example, CCSD calculations are coupled-cluster calculations involving excitations of single electrons (S), and pairs of electrons (double—D), while CCSDT calculations further include excitations of three electrons (triples—T). Møller–Plesset perturbation theory is based on adding a small perturbation (the correlation potential) to a zero-order Hamiltonian (the HF Hamiltonian, usually). In the Møller–Plesset perturbation theory approach, a number is used to indicate the order of the perturbation theory, so MP2 is the second-order theory and so on.

Another class of methods uses more than one Slater determinant as the reference wave function. The methods used to describe electron correlation within these calculations are similar in some ways to the methods listed above. These methods include multiconfigurational self-consistent field (MCSCF), multireference single and double configuration interaction (MRDCI), and N-electron valence state perturbation theory (NEVPT) methods.[§]

The classification of wave-function-based methods has two distinct components: the level of theory and the basis set. The level of theory defines the approximations that are introduced to describe electron–electron interactions. This is described by the array of acronyms introduced in the preceding paragraphs that describe various levels of theory. It has been suggested, only half-jokingly, that a useful rule for assessing the accuracy of a quantum chemistry calculation is that "the longer the acronym, the better the level of theory."[6] The second, and equally important, component in classifying wave-function-based methods is the basis set. In the simple example we gave in Section 1.6.1 of a spatially localized function, we formed an overall function by adding together three individual functions. If we were aiming to approximate a particular function in this way, for example, the solution of the Schrödinger

[§]This may be a good time to remind yourself that this overview of quantum chemistry is meant to act something like a phrase book or travel guide for a foreign city. Details of the methods listed here may be found in the Further Reading section at the end of this chapter.

equation, we could always achieve this task more accurately by using more functions in our sum. Using a basis set with more functions allows a more accurate representation of the true solution but also requires more computational effort since the numerical coefficients defining the magnitude of each function's contribution to the net function must be calculated. Just as there are multiple levels of theory that can be used, there are many possible ways to form basis sets.

To illustrate the role of the level of theory and the basis set, we will look at two properties of a molecule of CH_4, the C–H bond length and the ionization energy. Experimentally, the C–H bond length is 1.094 Å[7] and the ionization energy for methane is 12.61 eV. First, we list these quantities calculated with four different levels of theory using the same basis set in Table 1.1. Three of the levels of theory shown in this table are wave-function-based, namely HF, MP2, and CCSD. We also list results from a DFT calculation using the most popular DFT functional for isolated molecules, that is, the B3LYP functional. (We return at the end of this section to the characteristics of this functional.) The table also shows the computational time needed for each calculation normalized by the time for the HF calculation. An important observation from this column is that the computational time for the HF and DFT calculations are approximately the same—this is a quite general result. The higher levels of theory, particularly the CCSD calculation, take considerably more computational time than the HF (or DFT) calculations.

All of the levels of theory listed in Table 1.1 predict the C–H bond length with accuracy within 1%. One piece of cheering information from Table 1.1 is that the DFT method predicts this bond length as accurately as the much more computationally expensive CCSD approach. The error in the ionization energy predicted by HF is substantial, but all three of the other methods give better predictions. The higher levels of theory (MP2 and CCSD) give considerably more accurate results for this quantity than DFT.

Now we look at the properties of CH_4 predicted by a set of calculations in which the level of theory is fixed and the size of the basis set is varied.

TABLE 1.1 Computed Properties of CH₄ Molecule for Four Levels of Theory Using pVTZ Basis Set[a]

Level of Theory	C–H (Å)	Percent Error	Ionization (eV)	Percent Error	Relative Time
HF	1.085	−0.8	11.49	−8.9	1
DFT (B3LYP)	1.088	−0.5	12.46	−1.2	1
MP2	1.085	−0.8	12.58	−0.2	2
CCSD	1.088	−0.5	12.54	−0.5	18

[a]Errors are defined relative to the experimental value.

TABLE 1.2   Properties of CH_4 Calculated Using DFT (B3LYP) with Four Different Basis Sets[a]

Basis Set	Number of Basis Functions	C–H (Å)	Percent Error	Ionization (eV)	Percent Error	Relative Time
STO-3G	27	1.097	0.3	12.08	−4.2	1
cc-pVDZ	61	1.100	0.6	12.34	−2.2	1
cc-pVTZ	121	1.088	−0.5	12.46	−1.2	2
cc-pVQZ	240	1.088	−0.5	12.46	−1.2	13

[a]Errors are defined relative to the experimental value. Time is defined relative to the STO-3G calculation.

Table 1.2 contains results of this kind using DFT calculations with the B3LYP functional in each case. There is a complicated series of names associated with different basis sets. Without going into the details, let us just say that STO-3G is a very common "minimal" basis set while cc-pVDZ, cc-pVTZ, and cc-pVQZ (D stands for double, T for triple, etc.) is a popular series of basis sets that have been carefully developed to be numerically efficient for molecular calculations. The table lists the number of basis functions used in each calculation and also the computational time relative to the most rapid calculation. All of the basis sets listed in Table 1.2 give C–H bond lengths that are within 1% of the experimental value. The ionization energy, however, becomes significantly more accurate as the size of the basis set becomes larger.

One other interesting observation from Table 1.2 is that the results for the two largest basis sets, pVTZ and pVQZ, are identical (at least to the numerical precision we listed in the table). This occurs when the basis sets include enough functions to accurately describe the solution of the Schrödinger equation, and when it occurs the results are said to be "converged with respect to basis set." When it happens, this is a good thing! An unfortunate fact of nature is that a basis set that is large enough for one level of theory, say DFT, is not necessarily large enough for higher levels of theory. So the results in Table 1.2 do not imply that the pVTZ basis set used for the CCSD calculations in Table 1.1 were converged with respect to basis set.

In order to use wave-function-based methods to converge to the true solution of the Schrödinger equation, it is necessary to simultaneously use a high level of theory *and* a large basis set. Unfortunately, this approach is only feasible for calculations involving relatively small numbers of atoms because the computational expense associated with these calculations increases rapidly with the level of theory and the number of basis functions. For a basis set with N functions, for example, the computational expense of a conventional HF calculation typically requires $\sim N^4$ operations, while a conventional coupled-cluster calculation requires $\sim N^7$ operations. Advances have been made that improve the scaling of both HF and post-HF calculations. Even with these improvements, however you can appreciate the problem with scaling if you notice from Table 1.2 that a reasonable basis set for even a tiny molecule like CH_4 includes hundreds of basis functions. The computational expense of high-level wave-function-based methods means that these calculations are feasible for individual organic molecules containing 10–20 atoms, but physical systems larger than this fall into either the "very challenging" or "computationally infeasible" categories.

This brings our brief tour of quantum chemistry almost to an end. As the title of this book suggests, we are going to focus throughout the book on density functional theory calculations. Moreover, we will only consider methods based on spatially periodic functions—the so-called plane-wave methods. Plane-wave methods are the method of choice in almost all situations where the physical material of interest is an extended crystalline material rather than an isolated molecule. As we stated above, it is not appropriate to view methods based on periodic functions as "right" and methods based on spatially localized functions as "wrong" (or vice versa). In the long run, it will be a great advantage to you to understand both classes of methods since having access to a wide range of tools can only improve your chances of solving significant scientific problems. Nevertheless, if you are interested in applying computational methods to materials other than isolated molecules, then plane-wave DFT is an excellent place to start.

It is important for us to emphasize that DFT calculations can also be performed using spatially localized functions—the results in Tables 1.1 and 1.2 are examples of this kind of calculation. Perhaps the main difference between DFT calculations using periodic and spatially localized functions lies in the exchange–correlation functionals that are routinely used. In Section 1.4 we defined the exchange–correlation functional by what it does not include—it is the parts of the complete energy functional that are left once we separate out the contributions that can be written in simple ways. Our discussion of the HF method, however, indicates that it is possible to treat the exchange part of the problem in an exact way, at least in principle. The most commonly used functionals in DFT calculations based on spatially localized basis functions are "hybrid" functionals that mix the exact results for the exchange part of the functional with approximations for the correlation part. The B3LYP functional is by far the most widely used of these hybrid functionals. The B stands for Becke, who worked on the exchange part of the problem, the LYP stands for Lee, Yang, and Parr, who developed the correlation part of the functional, and the 3 describes the particular way that the results are mixed together. Unfortunately, the form of the exact exchange results mean that they can be efficiently implemented for applications based on spatially localized functions but not for applications using periodic functions! Because of this fact, the functionals that are commonly used in plane-wave DFT calculations do not include contributions from the exact exchange results.

1.7   WHAT CAN DFT NOT DO?

It is very important to come to grips with the fact that practical DFT calculations are *not* exact solutions of the full Schrödinger equation. This inexactness exists because the exact functional that the Hohenberg–Kohn theorem applies to is not known. So any time you (or anyone else) performs a DFT calculation, there is an intrinsic uncertainty that exists between the energies calculated with DFT and the true ground-state energies of the Schrödinger equation. In many situations, there is no direct way to estimate the magnitude of this uncertainty apart from careful comparisons with experimental measurements. As you read further through this book, we hope you will come to appreciate that there are many physical situations where the accuracy of DFT calculations is good enough to make powerful predictions about the properties of complex materials. The vignettes in Section 1.2 give several examples of this idea. We discuss the complicated issue of the accuracy of DFT calculations in Chapter 10.

There are some important situations for which DFT cannot be expected to be physically accurate. Below, we briefly discuss some of the most common problems that fall into this category. The first situation where DFT calculations have limited accuracy is in the calculation of electronic excited states. This can be understood in a general way by looking back at the statement of the Hohenberg–Kohn theorems in Section 1.4; these theorems only apply to the ground-state energy. It is certainly possible to make predictions about excited states from DFT calculations, but it is important to remember that these predictions are not—theoretically speaking—on the same footing as similar predictions made for ground-state properties.

A well-known inaccuracy in DFT is the underestimation of calculated band gaps in semiconducting and insulating materials. In isolated molecules, the energies that are accessible to individual electrons form a discrete set (usually described in terms of molecular orbitals). In crystalline materials, these energies must be described by continuous functions known as energy bands. The simplest definition of metals and insulators involves what energy levels are available to the electrons in the material with the highest energy once all the low-energy bands are filled in accordance with the Pauli exclusion principle. If the next available electronic state lies only at an infinitesimal energy above the highest occupied state, then the material is said to be a metal. If the next available electronic state sits a finite energy above the highest occupied state, then the material is not a metal and the energy difference between these two states is called the band gap. By convention, materials with "large" band gaps (i.e., band gaps of multiple electron volts) are called insulators while materials with "small" band gaps are called semiconductors. Standard DFT calculations with existing functionals have limited accuracy for band gaps, with errors larger than 1 eV being common when comparing with experimental data. A subtle feature of this issue is that it has been shown that even the formally exact Kohn–Sham exchange–correlation functional would suffer from the same underlying problem.[||]

Another situation where DFT calculations give inaccurate results is associated with the weak van der Waals (vdW) attractions that exist between atoms and molecules. To see that interactions like this exist, you only have to think about a simple molecule like CH_4 (methane). Methane becomes a liquid at sufficiently low temperatures and high enough pressures. The transportation of methane over long distances is far more economical in this liquid form than as a gas; this is the basis of the worldwide liquefied natural gas (LNG) industry. But to become a liquid, some attractive interactions between pairs of CH_4 molecules must exist. The attractive interactions are the van der Waals interactions, which, at the most fundamental level, occur because of correlations that exist between temporary fluctuations in the electron density of one molecule and the energy of the electrons in another molecule responding to these fluctuations. This description already hints at the reason that describing these interactions with DFT is challenging; van der Waals interactions are a direct result of long range electron correlation. To accurately calculate the strength of these interactions from quantum mechanics, it is necessary to use high-level wave-function-based methods that treat electron correlation in a systematic way. This has been done, for example, to calculate the very weak interactions that exist between pairs of H_2 molecules, where it is known experimentally that energy of two H_2 molecules in their most favored geometry is ~ 0.003 eV lower than the energy of the same molecules separated by a long distance.[8]

There is one more fundamental limitation of DFT that is crucial to appreciate, and it stems from the computational expense associated with solving the mathematical problem posed by DFT. It is reasonable to say that calculations that involve tens of atoms are now routine, calculations involving hundreds of atoms are feasible but are considered challenging research-level problems, and calculations involving a thousand or more atoms are possible but restricted to a small group of people developing state-of-the-art codes and using some of the world's largest computers. To keep this in a physical perspective, a droplet of water 1 μm in radius contains on the order of 10^{11} atoms. No conceivable increase in computing technology or code efficiency will allow DFT

[||]Development of methods related to DFT that can treat this situation accurately is an active area of research where considerable progress is being made. Two representative examples of this kind of work are P. Rinke, A. Qteish, J. Neugebauer, and M. Scheffler, Exciting Prospects for Solids: Exact-Exchange Based Functional Meet Quasiparticle Energy Calculations, *Phys. Stat. Sol.* **245** (2008), 929, and J. Uddin, J. E. Peralta, and G. E. Scuseria, Density Functional Theory Study of Bulk Platinum Monoxide, *Phys. Rev. B*, **71** (2005), 155112.

calculations to directly examine collections of atoms of this size. As a result, anyone using DFT calculations must clearly understand how information from calculations with extremely small numbers of atoms can be connected with information that is physically relevant to real materials.

1.8   DENSITY FUNCTIONAL THEORY IN OTHER FIELDS

For completeness, we need to point out that the name density functional theory is not solely applied to the type of quantum mechanics calculations we have described in this chapter. The idea of casting problems using functionals of density has also been used in the classical theory of fluid thermodynamics. In this case, the density of interest is the fluid density not the electron density, and the basic equation of interest is not the Schrödinger equation. Realizing that these two distinct scientific communities use the same name for their methods may save you some confusion if you find yourself in a seminar by a researcher from the other community.

1.9   HOW TO APPROACH THIS BOOK (REVISITED)

We began this chapter with an analogy about learning to drive to describe our aims for this book. Now that we have introduced much of the terminology associated with DFT and quantum chemistry calculations, we can state the subject matter and approach of the book more precisely. The remaining chapters focus on using plane-wave DFT calculations with commonly applied functionals to physical questions involving bulk materials, surfaces, nanoparticles, and molecules. Because codes to perform these plane-wave calculations are now widely available, we aim to introduce many of the issues associated with applying these methods to interesting scientific questions in a computationally efficient way.

The book has been written with two audiences in mind. The primary audience is readers who are entering a field of research where they will perform DFT calculations (and perhaps other kinds of computational chemistry or materials modeling) on a daily basis. If this describes you, it is important that you perform as many of the exercises at the end of the chapters as possible. These exercises have been chosen to require relatively modest computational resources while exploring most of the key ideas introduced in each chapter. Simply put, if your aim is to enter a field where you will perform calculations, then you must actually do calculations of your own, not just read about other people's work. As in almost every endeavor, there are many details that are best learned by experience. For readers in this group, we recommend reading through every chapter sequentially.

The second audience is people who are unlikely to routinely perform their own calculations, but who work in a field where DFT calculations have become a "standard" approach. For this group, it is important to understand the language used to describe DFT calculations and the strengths and limitations of DFT. This situation is no different from "standard" experimental techniques such as X-ray diffraction or scanning electron microscopy, where a working knowledge of the basic methods is indispensable to a huge community of researchers, regardless of whether they personally apply these methods. If you are in this audience, we hope that this book can help you become a sophisticated consumer of DFT results in a relatively efficient way. If you have a limited amount of time (a long plane flight, for example), we recommend that you read Chapter 3, Chapter 10, and then read whichever of Chapters 4–9 appears most relevant to you. If (when?) your flight is delayed, read one of the chapters that doesn't appear directly relevant to your specific research interests—we hope that you will learn something interesting.

We have consciously limited the length of the book in the belief that the prospect of reading and understanding an entire book of this length is more appealing than the alternative of facing (and carrying) something the size of a large city's phone book. Inevitably, this means that our coverage of various topics is limited in scope. In particular, we do not examine the details of DFT calculations using localized basis sets beyond the cursory treatment already presented in this chapter. We also do not delve deeply into the theory of DFT and the construction of functionals. In this context, the word "introduction" appears in the title of the book deliberately. You should view this book as an entry point into the vibrant world of DFT, computational chemistry, and materials modeling. By following the resources that are listed at the end of each chapter in the Further Reading section, we hope that you will continue to expand your horizons far beyond the introduction that this book gives.

We have opted to defer the crucial issue of the accuracy of DFT calculations until chapter 10, after introducing the application of DFT to a wide variety of physical properties in the preceding chapters. The discussion in that chapter emphasizes that this topic cannot be described in a simplistic way. Chapter 10 also points to some of the areas in which rapid developments are currently being made in the application of DFT to challenging physical problems.

REFERENCES

1. K. Honkala, A. Hellman, I. N. Remediakis, A. Logadottir, A. Carlsson, S. Dahl, C. H. Christensen, and J. K. Nørskov, Ammonia Synthesis from First-Principles Calculations, *Science* **307** (2005), 555.
2. R. Schweinfest, A. T. Paxton, and M. W. Finnis, Bismuth Embrittlement of Copper is an Atomic Size Effect, *Nature* **432** (2004), 1008.

3. K. Umemoto, R. M. Wentzcovitch, and P. B. Allen, Dissociation of $MgSiO_3$ in the Cores of Gas Giants and Terrestrial Exoplanets, *Science* **311** (2006), 983.

4. K. Umemoto, R. M. Wentzcovitch, D. J. Weidner, and J. B. Parise, $NaMgF_3$: A Low-Pressure Analog of $MgSiO_3$, *Geophys. Res. Lett.* **33** (2006), L15304.

5. W. Kohn, Nobel Lecture: Electronic Structure of Matter-Wave Functions and Density Functionals, *Rev. Mod. Phys.* **71** (1999), 1253.

6. C. J. Cramer and J. T. Roberts, Computational Analysis: Y_2K, *Science* **286** (1999), 2281.

7. G. Herzberg, *Electronic Spectra and Electronic Structure of Polyatomic Molecules*, Van Nostrand, New York, 1966.

8. P. Diep and J. K. Johnson, An Accurate H_2-H_2 Interaction Potential from First Principles, *J. Chem. Phys.* **112** (2000), 4465.

FURTHER READING

Throughout this book, we will list resources for further reading at the end of each chapter. You should think of these lists as pointers to help you learn about topics we have mentioned or simplified in a detailed way. We have made no attempt to make these lists exhaustive in any sense (to understand why, find out how many textbooks exist dealing with "quantum mechanics" in some form or another).

Among the many books on quantum mechanics that have been written, the following are good places to start if you would like to review the basic concepts we have touched on in this chapter:

P. W. Atkins and R. S. Friedman, *Molecular Quantum Mechanics*, Oxford University Press, Oxford, UK, 1997.

D. A. McQuarrie, *Quantum Chemistry*, University Science Books, Mill Valley, CA, 1983.

M. A. Ratner and G. C. Schatz, *Introduction to Quantum Mechanics in Chemistry*, Prentice Hall, Upper Saddle River, NJ, 2001.

J. Simons and J. Nichols, *Quantum Mechanics in Chemistry*, Oxford University Press, New York, 1997.

Detailed accounts of DFT are available in:

W. Koch and M. C. Holthausen, *A Chemist's Guide to Density Functional Theory*, Wiley-VCH, Weinheim, 2000.

R. M. Martin, *Electronic Structure: Basic Theory and Practical Methods*, Cambridge University Press, Cambridge, UK, 2004.

R. G. Parr and W. Yang, *Density-Functional Theory of Atoms and Molecules*, Oxford University Press, Oxford, UK, 1989.

Resources for learning about the wide range of quantum chemistry calculation methods that go beyond DFT include:

J. B. Foresman and A. Frisch, *Exploring Chemistry with Electronic Structure Methods*, Gaussian Inc., Pittsburgh, 1996.

A. Szabo and N. S. Ostlund, *Modern Quantum Chemistry: Introduction to Advanced Electronic Structure Theory*, Dover, Minneola, NY, 1996.

D. Young, *Computational Chemistry: A Practical Guide for Applying Techniques to Real World Problems*, Wiley, New York, 2001.

A book that gives a relatively brief overview of band theory is:

A. P. Sutton, *Electronic Structure of Materials*, Oxford University Press, Oxford, UK, 1993.

Two traditional sources for a more in-depth view of this topic are:

N. W. Ashcroft and N. D. Mermin, *Solid State Physics*, Saunders College Publishing, Orlando, 1976.

C. Kittel, *Introduction to Solid State Physics*, Wiley, New York, 1976.

A good source if you want to learn about the fluid thermodynamics version of DFT is:

H. Ted Davis, *Statistical Mechanics of Phases, Interfaces, and Thin Films*, Wiley-VCH, 1995.

2

DFT CALCULATIONS FOR SIMPLE SOLIDS

In this chapter, we explore how DFT calculations can be used to predict an important physical property of solids, namely their crystal structure. We will do this without delving into the technical details of actually performing such calculations because it is important to have a clear understanding of how DFT calculations can be used before getting too involved in the details of convergence and so on. This is not to say that these details are unimportant—they are absolutely crucial for performing reliable calculations! For the purposes of this chapter, we simply assume that we can use a DFT code to calculate the total energy of some collection of atoms. At the end of the chapter we give a few specific suggestions for doing calculations of this type that are needed if you perform the exercises that are suggested throughout the chapter.

2.1 PERIODIC STRUCTURES, SUPERCELLS, AND LATTICE PARAMETERS

It is quite likely that you are already familiar with the concept of crystal structures, but this idea is so central to using plane-wave DFT calculations that we are going to begin by considering the simplest possible example. Our task is to define the location of all atoms in a crystal of a pure metal. For now, imagine

Density Functional Theory: A Practical Introduction. By David S. Sholl and Janice A. Steckel
Copyright © 2009 John Wiley & Sons, Inc.

that we fill three-dimensional space with cubes of side length a and place a single metal atom at the corner of each cube. That is, the positions of all atoms are defined using normal three-dimensional Cartesian coordinates by $\mathbf{r} = (n_1 a, n_2 a, n_3 a)$ for any integers n_1, n_2, and n_3. This is called the simple cubic structure, and there is an element whose atoms are actually arranged this way in nature: polonium. What information do we need to specify to completely define the crystal structure of a simple cubic metal? Just one number, namely the lattice constant, a.

We now need to define a collection of atoms that can be used in a DFT calculation to represent a simple cubic material. Said more precisely, we need to specify a set of atoms so that when this set is repeated in every direction, it creates the full three-dimensional crystal structure. Although it is not really necessary for our initial example, it is useful to split this task into two parts. First, we define a volume that fills space when repeated in all directions. For the simple cubic metal, the obvious choice for this volume is a cube of side length a with a corner at $(0,0,0)$ and edges pointing along the x, y, and z coordinates in three-dimensional space. Second, we define the position(s) of the atom(s) that are included in this volume. With the cubic volume we just chose, the volume will contain just one atom and we could locate it at $(0,0,0)$. Together, these two choices have completely defined the crystal structure of an element with the simple cubic structure. The vectors that define the cell volume and the atom positions within the cell are collectively referred to as the *supercell*, and the definition of a supercell is the most basic input into a DFT calculation.

The choices we made above to define a simple cubic supercell are not the only possible choices. For example, we could have defined the supercell as a cube with side length $2a$ containing four atoms located at $(0,0,0)$, $(0,0,a)$, $(0,a,0)$, and $(a,0,0)$. Repeating this larger volume in space defines a simple cubic structure just as well as the smaller volume we looked at above. There is clearly something special about our first choice, however, since it contains the minimum number of atoms that can be used to fully define the structure (in this case, 1). The supercell with this conceptually appealing property is called the *primitive cell*.

If we go back to our definition of the primitive cell for the simple cubic structure, we are not required to place the atom in the supercell at $(0,0,0)$. We could just as well place the atom at $(a/2, a/2, a/2)$ or $(0,0,a/2)$ or even $(0.647a, 0.2293a, 0.184a)$. For every one of these choices, we are defining just one atom in the supercell, and the repetition of the supercell in space creates the simple cubic structure. Putting the atom at $(0,0,0)$ may seem like the best choice from an aesthetic point of view, although $(a/2, a/2, a/2)$ is also appealing because it puts the atom in the middle of the supercell. The point to remember is that any of these choices is mathematically equivalent, so they are all equally good from the point of view of doing an actual calculation.

You now know how to define a supercell for a DFT calculation for a material with the simple cubic crystal structure. We also said at the outset that we assume for the purposes of this chapter that we have a DFT code that can give us the total energy of some collection of atoms. How can we use calculations of this type to determine the lattice constant of our simple cubic metal that would be observed in nature? The sensible approach would be to calculate the total energy of our material as a function of the lattice constant, that is, $E_{tot}(a)$. A typical result from doing this type of calculation is shown in Fig. 2.1. The details of how these calculations (and the other calculations described in the rest of the chapter) were done are listed in the Appendix at the end of the chapter.

The shape of the curve in Fig. 2.1 is simple; it has a single minimum at a value of a we will call a_0. If the simple cubic metal exists with any value of a larger or smaller than a_0, the total energy of the material could be reduced by changing the lattice parameter to a_0. Since nature always seeks to minimize energy, we have made a direct physical prediction with our calculations: DFT predicts that the lattice parameter of our simple cubic material is a_0.

To extract a value of a_0 from our calculations, it is useful to think about the functional form of $E_{tot}(a)$. The simplest approach is to write the total energy using a truncated Taylor expansion:

$$E_{tot}(a) \cong E_{tot}(a_0) + \alpha(a - a_0) + \beta(a - a_0)^2 \qquad (2.1)$$

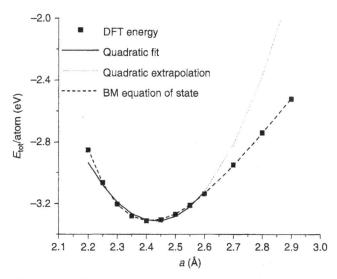

Figure 2.1 Total energy, E_{tot}, of Cu in the simple cubic crystal structure as a function of the lattice parameter, a. The filled symbols show the results of DFT calculations, while the three curves show the fits of the DFT data described in the text.

with $\alpha = dE_{tot}/da|_{a_0}$ and $\beta = \frac{1}{2}d^2E_{tot}/da^2|_{a_0}$. By definition, $\alpha = 0$ if a_0 is the lattice parameter corresponding to the minimum energy. This suggests that we can fit our numerical data to

$$E_{tot}(a) \cong E_0 + \beta(a - a_0)^2, \tag{2.2}$$

where E_0, β, and a_0 are treated as fitting parameters. The solid black curve shown in Fig. 2.1 is the result of fitting this curve to our data using values of a from 2.25 to 2.6 Å. This fitted curve predicts that a_0 is 2.43 Å.

Although we treated β simply as a fitting parameter in the equation above, it actually has direct physical significance. The equilibrium bulk modulus, B_0, of a material is defined by $B_0 = Vd^2E_{tot}/dV^2$, where the derivative is evaluated at the equilibrium lattice parameter. Comparing this expression with the Taylor expansion above, we find that $B_0 = \frac{2}{9}(1/a_0)\beta$. That is, the curve fitting we have performed gives us a value for both the equilibrium lattice parameter and the equilibrium bulk modulus. The bulk modulus from the solid curve in Fig. 2.1 is 0.641 eV/Å^3. These units are perfectly natural when doing a DFT calculation, but they are awkward for comparing to macroscopic data. Converting this result into more familiar units, we have predicted that $B_0 = 103$ GPa.

Our derivation of Eq. (2.2) gave a simple relationship between E_{tot} and a, but it is only valid for a small range of lattice constants around the equilibrium value. This can be seen directly from Fig. 2.1, where the quadratic fit to the data is shown as a gray curve for values of $a > 2.6$ Å. This range is of special interest because the DFT data in this range was not used in fitting the curve. It is clear from the figure that the fitted curve begins to deviate quite strongly from the DFT data as the lattice parameter increases. The root of this problem is that the overall shape of $E_{tot}(a)$ is not simply a quadratic function of the lattice parameter. More detailed mathematical treatments can give equations of state that relate these two quantities over a wider range of lattice constants. One well-known example is the Birch–Murnaghan equation of state for isotropic solids:

$$E_{tot}(a) = E_0 + \frac{9V_0B_0}{16}\left\{\left[\left(\frac{a_0}{a}\right)^2 - 1\right]^3 B_0'\right.$$

$$\left. + \left[\left(\frac{a_0}{a}\right)^2 - 1\right]^2\left[6 - 4\left(\frac{a_0}{a}\right)^2\right]\right\}. \tag{2.3}$$

In this expression, a_0 is the equilibrium lattice constant, V_0 is the equilibrium volume per atom, B_0 is the bulk modulus at zero pressure, $P = 0$, and $B_0' = (\partial B/\partial P)_T$. To apply this equation to our data, we treat a_0, B_0, and B_0' and E_0 as fitting parameters. The results of fitting this equation of state to the full set of DFT data shown in Fig. 2.1 are shown in the figure with a dashed line. It is clear from the figure that the equation of state allows us to accurately fit the data. The outcome from this calculation is the prediction that for Cu in a cubic crystal structure, a_0 is 2.41 Å and $B_0 = 102$ GPa. It should not surprise you that these predictions are very similar to the ones we made with the simpler quadratic model for $E_{tot}(a)$ because for lattice parameters close to a_0, Eq. (2.3) reduces to Eq. (2.2).

2.2 FACE-CENTERED CUBIC MATERIALS

The simple cubic crystal structure we discussed above is the simplest crystal structure to visualize, but it is of limited practical interest at least for elements in their bulk form because other than polonium no elements exist with this structure. A much more common crystal structure in the periodic table is the face-centered-cubic (fcc) structure. We can form this structure by filling space with cubes of side length a that have atoms at the corners of each cube and also atoms in the center of each face of each cube. We can define a supercell for an fcc material using the same cube of side length a that we used for the simple cubic material and placing atoms at $(0,0,0)$, $(0,a/2,a/2)$, $(a/2,0,a/2)$, and $(a/2,a/2,0)$. You should be able to check this statement for yourself by sketching the structure.

Unlike the definition of the supercell for a simple cubic material we used above, this supercell for an fcc metal contains four distinct atoms. This gives us a hint that we may not have described the primitive cell. This suspicion can be given more weight by drawing a sketch to count the number of atoms that neighbor an atom at a corner in our "natural" structure and then repeating this exercise for an atom in a cube face in our structure. Doing this shows that both types of atoms have 12 neighboring atoms arranged in the same geometry and that the distances between any atom and all of its 12 neighbors are identical.

The primitive cell for the fcc metal can be defined by connecting the atom at the origin in the structure defined above with three atoms in the cube faces adjacent to that atom. That is, we define cell vectors

$$\mathbf{a}_1 = a\left(\tfrac{1}{2},\tfrac{1}{2},0\right) \qquad \mathbf{a}_2 = a\left(0,\tfrac{1}{2},\tfrac{1}{2}\right) \qquad \mathbf{a}_3 = a\left(\tfrac{1}{2},0,\tfrac{1}{2}\right). \tag{2.4}$$

These vectors define the fcc lattice if we place atoms at positions

$$\mathbf{r} = n_1\mathbf{a}_1 + n_2\mathbf{a}_2 + n_3\mathbf{a}_3 \tag{2.5}$$

for all integers n_1, n_2, and n_3. This situation is illustrated in Fig. 2.2. We can see from these definitions that there is one atom in the cell, so it must be the primitive cell. Also, the geometry of the crystal is fully defined by a single

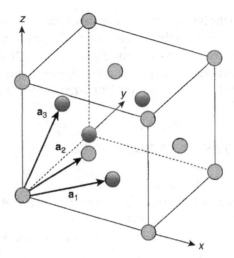

Figure 2.2   Illustration of the cell vectors of the fcc metal defined in Eq. (2.4).

parameter, a. The distance between nearest-neighbor atoms in an fcc metal is $a/\sqrt{2}$. A third important observation is that the cell vectors are not orthogonal (i.e., none of $\mathbf{a}_i \cdot \mathbf{a}_j$ are zero).

The results from calculating the total energy of fcc Cu as a function of the lattice parameter are shown in Fig. 2.3. The shape of the curve is similar to the one we saw for Cu in the simple cubic crystal structure (Fig. 2.1), but

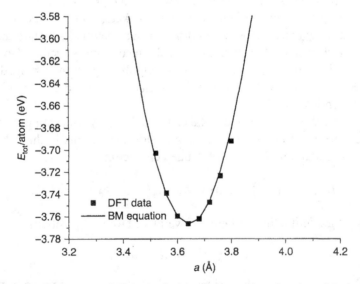

Figure 2.3   Total energy, E, of Cu in the fcc crystal structure as a function of the lattice parameter, a. Data points are from DFT calculations and the dashed curve is the Birch–Murnaghan equation of state.

the minimum energy in the fcc structure has a lower value than the minimum energy for the simple cubic crystal. This observation has a simple physical interpretation: the fcc crystal structure of Cu is more stable than the simple cubic crystal structure. This conclusion is not surprising since Cu is in reality an fcc metal, but it is pleasing to find that our DFT calculations are in agreement with physical reality. We can also compare our calculated lattice constant with the experimental result. From the curve in Fig. 2.3 (the Birch–Murnaghan fit to the DFT data) we predict that $a_0 = 3.64$ Å and $B_0 = 142$ GPa for Cu. Experimentally, the Cu lattice constant is 3.62 Å and $B_0 = 140$ GPa. For both quantities, our predictions are very close (although not equal to) the experimental values.

2.3 HEXAGONAL CLOSE-PACKED MATERIALS

It is not hard to understand why many metals favor an fcc crystal structure: there is no packing of hard spheres in space that creates a higher density than the fcc structure. (A mathematical proof of this fact, known as the Kepler conjecture, has only been discovered in the past few years.) There is, however, one other packing that has exactly the same density as the fcc packing, namely the hexagonal close-packed (hcp) structure. As our third example of applying DFT to a periodic crystal structure, we will now consider the hcp metals.

The supercell for an hcp metal is a little more complicated than for the simple cubic or fcc examples with which we have already dealt. The supercell can be defined using the following cell vectors:

$$\mathbf{a}_1 = (a,0,0) \qquad \mathbf{a}_2 = \left(\frac{a}{2}, \frac{\sqrt{3}a}{2}, 0\right) \qquad \mathbf{a}_3 = (0,0,c) \qquad (2.6)$$

and placing two distinct atoms within the supercell with Cartesian coordinates $(0,0,0)$ and $(a/2, a/2\sqrt{3}, c/2)$. You should convince yourself that if you consider only the vectors $\mathbf{a}_1$ and $\mathbf{a}_2$ and the atom at $(0,0,0)$ in the supercell these together generate a hexagonal array of atoms in the x–y plane. Also notice that this definition of the complete supercell involves two unknown parameters, a and c, rather than the one unknown lattice constant that we had for the simple cubic and fcc structures. In the ideal hard sphere packing, $c = \sqrt{8/3}\, a = 1.633a$, and the distance between all possible pairs of adjacent atoms in the crystal are identical. In real hcp metals, however, small distortions of the crystal are often observed. For example, for scandium, $c/a = 1.59$ experimentally.

The definition of the hcp supercell given above is useful to introduce one more concept that is commonly used in defining atomic coordinates in periodic geometries. As our definition stands, the vectors defining the shape of the supercell (the lattice vectors) and the atom positions have been given in three-dimensional Cartesian coordinates. It is typically more convenient to define the lattice vectors in Cartesian coordinates and then to define the atom positions in terms of the lattice vectors. In general, we can write the position of atom j in the supercell as

$$\mathbf{r}_j = \sum_{j=1}^{3} f_{j,i}\mathbf{a}_i. \tag{2.7}$$

Because we can always choose each atom so it lies within the supercell, $0 \leq f_{j,i} \leq 1$ for all i and j. These coefficients are called the *fractional coordinates* of the atoms in the supercell. The fractional coordinates are often written in terms of a vector for each distinct atom. In the hcp structure defined above, for example, the two atoms lie at fractional coordinates $(0,0,0)$ and $(\frac{1}{3},\frac{1}{3},\frac{1}{3})$. Notice that with this definition the only place that the lattice parameters appear in the definition of the supercell is in the lattice vectors. The definition of a supercell with a set of lattice vectors and a set of fractional coordinates is by far the most convenient way to describe an arbitrary supercell, and it is the notation we will use throughout the remainder of this book. Most, if not all, popular DFT packages allow or require you to define supercells using this notation.

Using DFT to predict the lattice constant of Cu in the simple cubic or fcc crystal structures was straightforward; we just did a series of calculations of the total energy as a function of the lattice parameter, a. The fact that the hcp structure has two independent parameters, a and c, complicates this process. Most DFT packages have the capability to handle multivariable problems like this in an automated way, but for now let us stick with the assumption that we only know how to use our package to compute a total energy for one cell volume and geometry at a time. One way to proceed is to simply fix the value of c/a and then calculate a series of total energies as a function of a. Results from a series of calculations like this are shown in Fig. 2.4. In this figure, the lines simply connect the data points to guide the eye. From these calculations, we see that the distortion along the c axis away from hard sphere packing for Cu is predicted to be small. More importantly, the minimum energy of the hcp Cu structure is larger than the minimum energy for the fcc structure by $\sim 0.015\,\mathrm{eV/atom}$ (cf. Fig. 2.3), so our calculations agree with the observation that Cu is an fcc metal, not an hcp metal. The predicted energy difference, $1.4\,\mathrm{kJ/mol}$, between the two crystal structures is quite small. This is not unreasonable since the two structures are very similar in many respects.

There are (at least) two things that our calculations for hcp Cu should make you think about. The first concerns the numerical accuracy of our DFT calculations. Can we reliably use these calculations to distinguish between the

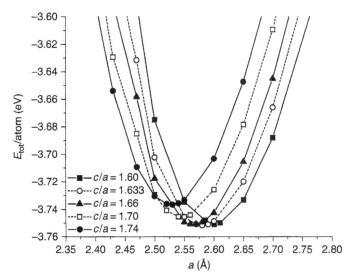

Figure 2.4   Total energy, E_{tot}, of Cu in the hcp crystal structure for several values of c/a. Each data point is from a DFT calculation. Straight lines connecting the data points are to guide the eye.

stability of two crystal structures that differ by only $\sim 1\ kJ/mol$ in energy? What about structures that differ by $\sim 0.1\ kJ/mol$ in energy? The answer to this question is intimately tied to our ability to numerically solve the complicated mathematical problem defined by DFT for a particular set of atoms. We will be occupied by this crucial topic of numerical convergence for much of the next chapter.

Second, it should be clear that performing the calculations for an example where we had to determine two lattice parameters was considerably more work than when we only had to determine one. This example is partly intended to give you a taste for the fact that when there are many degrees of freedom in a problem, whether they are atom positions within a supercell or the lattice parameters associated with cell vectors, minimizing the overall energy by varying one parameter at a time soon becomes untenable. After discussing numerical convergence, the next chapter deals with the important concept of how energy minima can be efficiently calculated for complex configurations of atoms without having to systematically vary each degree of freedom.

2.4 CRYSTAL STRUCTURE PREDICTION

It is tempting to say that we have predicted that crystal structure of Cu with our calculations in the previous sections, but this is not strictly true. To be precise, we should say that we have predicted that fcc Cu is more stable than hcp Cu or simple cubic Cu. Based on our calculations alone we cannot exclude the possibility that Cu in fact adopts some other crystal structure that we have not examined. For Cu this is a fairly pedantic point since we already know from experiments that it is an fcc metal. So we could state that our calculations are entirely consistent with the experimental crystal structure and that we now have a prediction for the lattice parameter of fcc Cu.

To make this point in another way, imagine that you have been asked to predict the crystal structure of di-yttrium potassium (Y_2K), a substance for which no experimental data is available. You could attempt this task using DFT by making a list of all known crystal structures with stoichiometry AB_2, then minimizing the total energy of Y_2K in each of these crystal structures. This is far from a simple task; more than 80 distinct AB_2 crystal structures are known, and many of them are quite complicated. To give just one example, $NiMg_2$, $NbZn_2$, $ScFe_2$, $ThMg_2$, $HfCr_2$, and UPt_2 all exist in an ordered structure known as the C36 hexagonal Laves phase that has 106 distinct atoms in the primitive cell. Even if you completed the somewhat heroic task of performing all these calculations, you could not be sure that Y_2K does not actually form a new crystal structure that has not been previously observed! This discussion illustrates why determining the crystal structure of new compounds remains an interesting scientific endeavor. The main message from this discussion is that DFT is very well suited to predicting the energy of crystal structures within a set of potential structures, but calculations alone are almost never sufficient to truly predict new structures in the absence of experimental data.

2.5 PHASE TRANSFORMATIONS

All of our analysis of the Cu crystal structure has been based on the reasonable idea that the crystal structure with the lowest energy is the structure preferred by nature. This idea is correct, but we need to be careful about how we define a materials' energy to make it precise. To be precise, the preferred crystal structure is the one with the lowest Gibbs free energy, $G = G(P, T)$. The Gibbs free energy can be written as

$$G(P, T) = E_{coh} + PV - TS, \qquad (2.8)$$

where E_{coh}, V, and S are the cohesive energy (i.e., the energy to pull a material apart into a collection of isolated atoms), volume, and entropy of a material. If we are comparing two possible crystal structures, then we are interested in the change in Gibbs free energy between the two structures:

$$\Delta G(P, T) = \Delta E_{coh} + P\,\Delta V - T\,\Delta S. \qquad (2.9)$$

In solids, the first two terms tend to be much larger than the entropic contribution from the last term in this expression, so

$$\Delta G(P, T) \cong \Delta E_{coh} + P \Delta V. \tag{2.10}$$

Both of the quantities on the right-hand side have a simple meaning in terms of the energies we have been calculating with DFT. The change in cohesive energy between the two structures is just the difference between the DFT total energies. The pressure that is relevant for any point on the energy curves we have calculated is defined by

$$P = -\frac{\partial E_{coh}}{\partial V}. \tag{2.11}$$

In Sections 2.1–2.3 we interpreted the minimum in a plot of the DFT energy as a function of the lattice parameter as the preferred lattice parameter for the crystal structure used in the calculations. Looking at Eqs. (2.8) and (2.11), you can see that a more precise interpretation is that a minimum of this kind defines the preferred lattice parameter at $P = 0$ and $T = 0$.

An interesting consequence of Eq. (2.10) is that two crystal structures with different cohesive energies can have the same Gibbs free energy if $\Delta E_{coh} = -P \Delta V$. Comparing this condition with Eq. (2.11), you can see that two structures satisfy this condition if they share a common tangent on a plot of ΔE_{coh} as a function of V. This situation is illustrated in Fig. 2.5. In this figure, the preferred crystal structure at $P = 0$ is structure 1, and the lattice parameter of this preferred structure defines a volume V_0. Moving along the

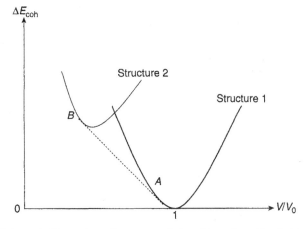

Figure 2.5   Schematic illustration of a pressure-induced transformation between two crystal structures of a material.

curve for the cohesive energy from the minimum at $V/V_0 = 1$ toward the point labeled A is equivalent to increasing the pressure on the material. When the pressure corresponding to point A is reached, the curve for structure 1 shares a common tangent with structure 2. At higher pressures, the Gibbs free energy of structure 2 is lower than for structure 1, so the figure predicts that a pressure-induced phase transformation would occur at this pressure that would change the crystal structure of the material. In this phase change the cohesive energy goes up, but this change is balanced within the free energy by the reduction in the volume of the material.

Pressure-induced phase transformations are known to occur for a wide range of solids. Bulk Si, for example, has the diamond structure at ambient conditions but converts to the β-tin structure at pressures around 100 kbar. Figure 2.5 shows how it is possible to use the kinds of information we have calculated in this chapter using DFT to predict the existence of pressure-induced phase transformations. It was essentially this idea that was used to make the geologically relevant predictions of the properties of minerals such as $MgSiO_3$ that were mentioned as one of the motivating examples in Section 1.2.

EXERCISES

To perform these exercises, you will need access to a code that performs plane-wave DFT calculations and be able to use this code to determine the total energy of a supercell. As you will discover in subsequent chapters, there are a number of input parameters that you need to provide to any DFT code. At this point, we suggest that you attempt the exercises below using, where possible, default input parameters, perhaps set up for you by a more experienced colleague. This should make you curious about what these input parameters mean, a curiosity that should be satisfied in the following chapters. The aim of the following exercises is to develop some proficiency with setting up supercells for bulk materials and practice some of the issues associated with predicting the crystal structures of these materials.

1. Perform calculations to determine whether Pt prefers the simple cubic, fcc, or hcp crystal structure. Compare your DFT-predicted lattice parameter(s) of the preferred structure with experimental observations.

2. Hf is experimentally observed to be an hcp metal with $c/a = 1.58$. Perform calculations to predict the lattice parameters for Hf and compare them with experimental observations.

3. A large number of solids with stoichiometry AB form the CsCl structure. In this structure, atoms of A define a simple cubic structure and atoms of B reside in the center of each cube of A atoms. Define the cell vectors

and fractional coordinates for the CsCl structure, then use this structure to predict the lattice constant of ScAl.

4. Another common structure for AB compounds is the NaCl structure. In this structure, A and B atoms alternate along any axis of the simple cubic structure. Predict the lattice parameter for ScAl in this structure and show by comparison to your results from the previous exercise that ScAl does not prefer the NaCl structure.

FURTHER READING

The fundamental concepts and notations associated with crystal structures are described in almost all textbooks dealing with solid-state physics or materials science. Two well-known examples are C. Kittel, *Introduction to Solid State Physics*, Wiley, New York, 1976, and N. W. Ashcroft and N. D. Mermin, *Solid State Physics*, Saunders College, Orlando, FL, 1976.

For a detailed account of the crystal structures of AB compounds, as well as an excellent overview of the various notations used to define crystal structures, see David Pettifor *Bonding and Structure of Molecules and Solids*, Oxford University Press, Oxford, UK, 1995.

For an interesting history of the Kepler conjecture, see G. G. Szpiro, *Kepler's Conjecture: How Some of the Greatest Minds in History Helped Solve One of the Oldest Math Problems in the World*, Wiley, Hoboken, NJ, 2003.

To see how quantum chemistry was used to predict the gas-phase properties of Y_2K, see C. J. Cramer and J. T. Roberts, *Science* **286** (1999), 2281. To appreciate the timeliness of this publication, note the date of publication.

APPENDIX CALCULATION DETAILS

In each chapter where we give results from specific calculations, we will include an appendix like this one where some details about how these calculations were performed are listed.

All of the calculations we describe throughout the book were done using the Vienna *ab initio* Simulation Package (VASP).* Although we enjoy using VASP and have the greatest appreciation for the people who have put many person-years of effort into developing it, our use of it here is not intended to be a formal endorsement of this package. VASP is one of several widely used plane-wave DFT packages, and other equally popular software packages could have been used for all the calculations we describe. The choice of which package is right for you will be influenced by factors such as the availability of

---

*Information about this package is available from http://cms.mpi.univie.ac.at/~vasp.

licenses for the software at your institution and the experience your and your co-workers have accumulated with one or more packages. We have attempted to only use features of VASP that are also available in essentially all plane-wave DFT codes, so you should have little trouble reproducing the calculations we have shown as examples using any mainstream code.

Unless otherwise stated, all of our calculations used the generalized gradient approximation as defined by the Perdew–Wang 91 functional. In shorthand, we used PW91-GGA calculations. Unless otherwise stated k points were placed in reciprocal space using the Monkhorst–Pack method. Some specific details for the calculations from each section of the chapter are listed below.

Section 2.1   The cubic Cu calculations in Fig. 2.1 used a cubic supercell with 1 Cu atom, a cutoff energy of 292 eV, and $12 \times 12 \times 12$ k points.

Section 2.2   The fcc Cu calculations in Fig. 2.3 used a cubic supercell with 4 Cu atoms, a cutoff energy of 292 eV, and $12 \times 12 \times 12$ k points.

Section 2.3   The hcp Cu calculations in Fig. 2.4 used a primitive supercell with 2 Cu atoms, a cutoff energy of 292 eV, and $12 \times 12 \times 8$ k points placed in reciprocal space using a Γ-centered grid.

3

NUTS AND BOLTS OF DFT CALCULATIONS

Throughout Chapter 2, we deliberately ignored many of the details that are necessary in performing DFT calculations in order to illustrate some of the physical quantities that these calculations can treat. This state of affairs is a useful starting point when you are learning about these methods, but it is, of course, not a useful strategy in the longer term. In this chapter, we dig into some of the details that make the difference between DFT calculations that provide reliable physical information and calculations that simply occupy time on your computer.

A key concept that we will reiterate many times is *convergence*. As you perform DFT calculations (or as you interact with other people who are performing such calculations), you should always be asking whether the calculations are *converged*.* What do we mean by convergence? To answer this question, it is useful to briefly look back at the description of DFT given in Chapter 1. The ground-state electron density of a configuration of atoms as defined by DFT is defined by the solution to a complicated set of mathematical equations. To actually solve this problem on a computer, we must make a series of numerical approximations: integrals in multidimensional space must be evaluated by

---

*This forms the basis for a foolproof question that can be asked of any theoretician when they give a talk: "Can you comment on how well converged your results are?" There is an equivalent generic question for experimenters: "Are there possible effects of contamination in your experiments?"

Density Functional Theory: A Practical Introduction. By David S. Sholl and Janice A. Steckel
Copyright © 2009 John Wiley & Sons, Inc.

examining the function to be integrated at a finite collection of points, solutions that formally are expressed as infinite sums must be truncated to finite sums, and so on. In each numerical approximation of this kind, it is possible to find a solution that is closer and closer to the exact solution by using more and more computational resources. This is the process we will refer to as convergence. A "well-converged" calculation is one in which the numerically derived solution accurately approximates the true solution of the mathematical problem posed by DFT with a specific exchange–correlation functional.

The concept of numerical convergence is quite separate from the question of whether DFT accurately describes physical reality. The mathematical problem defined by DFT is not identical to the full Schrödinger equation (because we do not know the precise form of the exchange–correlation functional). This means that the exact solution of a DFT problem is not identical to the exact solution of the Schrödinger equation, and it is the latter that we are presumably most interested in. This issue, the physical accuracy of DFT, is of the utmost important, but it is more complicated to fully address than the topic of numerical convergence. The issue of the physical accuracy of DFT calculations is addressed in Chapter 10.

In this chapter, we first concentrate on what is required to perform well-converged DFT calculations. After all, we need to be able to confidently find precise solutions to the numerical problems defined by DFT before we can reasonably discuss the agreement (or lack thereof) between DFT results and physical reality.

3.1   RECIPROCAL SPACE AND k POINTS

Our first foray into the realm of numerical convergence takes us away from the comfortable three-dimensional physical space where atom positions are defined and into what is known as reciprocal space. The concepts associated with reciprocal space are fundamental to much of solid-state physics; that there are many physicists who can barely fathom the possibility that anyone might find them slightly mysterious. It is not our aim here to give a complete description of these concepts. Several standard solid-state physics texts that cover these topics in great detail are listed at the end of the chapter. Here, we aim to cover what we think are the most critical ideas related to how reciprocal space comes into practical DFT calculations, with particular emphasis on the relationship between these ideas and numerical convergence.

3.1.1   Plane Waves and the Brillouin Zone

We emphasized in Chapter 2 that we are interested in applying DFT calculations to arrangements of atoms that are periodic in space. We defined the shape of the cell that is repeated periodically in space, the supercell, by lattice vectors $\mathbf{a}_1$, $\mathbf{a}_2$, and $\mathbf{a}_3$. If we solve the Schrödinger equation for this periodic system, the solution must satisfy a fundamental property known as Bloch's theorem, which states that the solution can be expressed as a sum of terms with the form

$$\phi_{\mathbf{k}}(\mathbf{r}) = \exp(i\mathbf{k} \cdot \mathbf{r}) u_{\mathbf{k}}(\mathbf{r}), \tag{3.1}$$

where $u_{\mathbf{k}}(\mathbf{r})$ is periodic in space with the same periodicity as the supercell. That is, $u_{\mathbf{k}}(\mathbf{r} + n_1\mathbf{a}_1 + n_2\mathbf{a}_2 + n_3\mathbf{a}_3) = u_{\mathbf{k}}(\mathbf{r})$ for any integers n_1, n_2, and n_3. This theorem means that it is possible to try and solve the Schrödinger equation for each value of $\mathbf{k}$ independently. We have stated this result in terms of solutions of the Schrödinger equation, but it also applies to quantities derived from solutions to this equation such as the electron density.

It turns out that many parts of the mathematical problems posed by DFT are much more convenient to solve in terms of $\mathbf{k}$ than they are to solve in terms of $\mathbf{r}$. Because the functions $\exp(i\mathbf{k} \cdot \mathbf{r})$ are called plane waves, calculations based on this idea are frequently referred to as plane-wave calculations. The space of vectors $\mathbf{r}$ is called real space, and the space of vectors $\mathbf{k}$ is called reciprocal space (or simply k space). The idea of using reciprocal space is so central to the calculations we will discuss in the rest of the book that it is important to introduce several features of k space.

Just as we defined positions in real space in terms of the lattice vectors $\mathbf{a}_1$, $\mathbf{a}_2$, and $\mathbf{a}_3$, it is useful to define three vectors that define positions in reciprocal space. These vectors are called the reciprocal lattice vectors, $\mathbf{b}_1$, $\mathbf{b}_2$, and $\mathbf{b}_3$, and are defined so that $\mathbf{a}_i \cdot \mathbf{b}_j$ is 2π if $i = j$ and 0 otherwise. This choice means that

$$\mathbf{b}_1 = 2\pi \frac{\mathbf{a}_2 \times \mathbf{a}_3}{\mathbf{a}_1 \cdot (\mathbf{a}_2 \times \mathbf{a}_3)}, \quad \mathbf{b}_2 = 2\pi \frac{\mathbf{a}_3 \times \mathbf{a}_1}{\mathbf{a}_2 \cdot (\mathbf{a}_3 \times \mathbf{a}_1)}, \quad \mathbf{b}_3 = 2\pi \frac{\mathbf{a}_1 \times \mathbf{a}_2}{\mathbf{a}_3 \cdot (\mathbf{a}_1 \times \mathbf{a}_2)}. \tag{3.2}$$

A simple example of this calculation is the simple cubic lattice we discussed in Chapter 2. In that case, the natural choice for the real space lattice vectors has $|\mathbf{a}_i| = a$ for all i. You should verify that this means that the reciprocal lattice vectors satisfy $|\mathbf{b}_i| = 2\pi/a$ for all i. For this system, the lattice vectors and the reciprocal lattice vectors both define cubes, the former with a side length of a and the latter with a side length of $2\pi/a$.

In Chapter 2 we mentioned that a simple cubic supercell can be defined with lattice vectors $\mathbf{a}_i = a$ or alternatively with lattice vectors $\mathbf{a}_i = 2a$. The first choice uses one atom per supercell and is the primitive cell for the simple cubic material, while the second choice uses eight atoms per supercell. Both choices define the same material. If we made the second choice, then our reciprocal lattice vectors would also change—they would become $\mathbf{b}_i = \pi/a$. This example illustrates an important general observation: *larger lattice vectors in real space correspond to shorter lattice vectors in reciprocal space.*

The three-dimensional shape defined by the reciprocal lattice vectors is not always the same as the shape of the supercell in real space. For the fcc primitive cell, we showed in Chapter 2 that

$$\mathbf{a}_1 = a\left(\tfrac{1}{2},\tfrac{1}{2},0\right), \quad \mathbf{a}_2 = a\left(0,\tfrac{1}{2},\tfrac{1}{2}\right), \quad \mathbf{a}_3 = a\left(\tfrac{1}{2},0,\tfrac{1}{2}\right). \tag{3.3}$$

The reciprocal vectors in this case are

$$\mathbf{b}_1 = \frac{2\pi}{a}(1,1,-1), \quad \mathbf{b}_2 = \frac{2\pi}{a}(-1,1,1), \quad \mathbf{b}_3 = \frac{2\pi}{a}(1,-1,1). \tag{3.4}$$

Again, notice that the length of the reciprocal lattice vectors are inversely related to the reciprocal of the length of the real space lattice vectors. Views of these lattice vectors in real space and reciprocal space are shown in Fig. 3.1.

We previously introduced the concept of a primitive cell as being the super-cell that contains the minimum number of atoms necessary to fully define a periodic material with infinite extent. A more general way of thinking about the primitive cell is that it is a cell that is minimal in terms of volume but still contains all the information we need. This concept can be made more precise by considering the so-called Wigner–Seitz cell. We will not go into

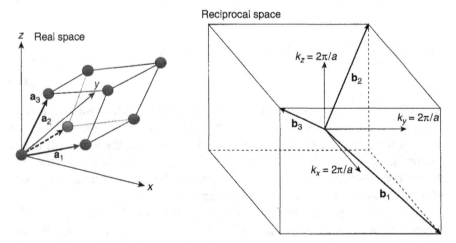

Figure 3.1   View of the real space and reciprocal space lattice vectors for the fcc primitive cell. In the real space picture, circles represent atoms. In the reciprocal space picture, the basis vectors are shown inside a cube with side length $4\pi/a$ centered at the origin.

the details of this construction here but simply state that the Wigner–Seitz cell can be defined for reciprocal lattice vectors just as easily as it can be for real space vectors. In plainer language, we can define a primitive cell in reciprocal space. Because this cell has many special properties, it is given a name: it is the *Brillouin zone* (often abbreviated to BZ). The Brillouin zone plays a central role in the band theory of materials. Several points in the Brillouin zone with special significance are given individual names. The most important of these is the point where $\mathbf{k} = 0$; this location in k space is called the Γ point. As a final example of the link between length in real space and reciprocal space, we note that the volume of the BZ, V_{BZ}, and the volume of the primitive cell in real space defined by the Wigner–Seitz construction, V_{cell}, are related by

$$V_{BZ} = \frac{(2\pi)^3}{V_{cell}}. \tag{3.5}$$

3.1.2 Integrals in k Space

Why is the Brillouin zone so important in plane-wave DFT calculations? The simple answer is that in a practical DFT calculation, a great deal of the work reduces to evaluating integrals of the form

$$\bar{g} = \frac{V_{cell}}{(2\pi)^3} \int_{BZ} g(\mathbf{k}) \, d\mathbf{k}. \tag{3.6}$$

The key features of this integral are that it is defined in reciprocal space and that it integrates only over the possible values of $\mathbf{k}$ in the Brillouin zone. Rather than examining in detail where integrals such as these come from, let us consider instead how we can evaluate them numerically.

Before we try and evaluate integrals such as those in Eq. (3.6), we will look at the simpler task of evaluating $\int_{-1}^{1} f(x) \, dx$ numerically. You hopefully remember from calculus that this integral can be thought of as the area under the curve defined by $f(x)$ on the interval $[-1,1]$. This interpretation suggests that a simple way to approximate the integral is to break up the interval into pieces of equal size and estimate the area under the curve by treating the curve as a straight line between the end points of the interval. This gives us the trapezoidal method:

$$\int_{-1}^{1} f(x) \, dx \cong \frac{h}{2} \left[f(-1) + f(+1) + 2\sum_{j=1}^{n-1} f(x_j) \right] \tag{3.7}$$

with $x_j = -1 + jh$ and $h = 2/n$.

TABLE 3.1   Approximations to the Integral $\int_{-1}^{1} \frac{\pi x}{2} \sin(\pi x)\, dx = 1$ Using the Trapezoidal and Legendre Quadrature Methods

N	Trapezoidal Method	Legendre Quadrature Method
2	0.6046	1.7605
3	0.7854	0.8793
4	0.8648	1.0080
5	0.9070	0.9997

As a simple test case, we can evaluate $\int_{-1}^{1} (\pi x/2) \sin(\pi x)\, dx$. This integral can be evaluated exactly via integration by parts, and we have chosen the function so that the value of the integral is exactly 1. Some results from applying the trapezoidal method to our test integral are shown in Table 3.1. Not surprisingly, as we use a smaller step size for the space between values of x where we evaluate the function, the method becomes more accurate. The results in Table 3.1 suggest that using a value of n much larger than 5 would be necessary to evaluate the integral to, say, an accuracy of 1%.

Two features of the trapezoidal method are that we use a uniform spacing between the positions where we evaluate $f(x)$ and that every evaluation of $f(x)$ (except the end points) is given equal weight. Neither of these conditions is necessary or even desirable. An elegant class of integration methods called Gaussian quadrature defines methods that have the form

$$\int_{-1}^{1} f(x)\, dx \cong \sum_{j=1}^{n} c_j f(x_j), \tag{3.8}$$

where the integration points x_j are related to roots of orthogonal polynomials and the weights c_j are related to integrals involving these polynomials. For integrals on the domain $[-1,1]$, this approach is called Legendre quadrature. To give one specific example, when $n = 3$, the weights and integration points in Eq. (3.8) are $x_1 = -x_3 = 0.775967$, $x_2 = 0$, $c_1 = -c_3 = 0.555555$, and $c_2 = 0.88888$. The results from applying this method to our test integral are listed in Table 3.1. In striking contrast to the trapezoidal method, the results converge very quickly to the correct result as n is increased. In this case, the error in the integral is $<1\%$ once $n > 3$.

The example above of numerically integrating a one-dimensional function can be summarized in three main points that also apply to multidimensional integrals:

1. Integrals can be approximated by evaluating the function to be integrated at a set of discrete points and summing the function values with appropriate weighting for each point.

2. Well-behaved numerical methods of this type will give more accurate results as the number of discrete points used in the sum is made larger. In the limit of using a very large number of points, these numerical methods converge to the exact result for the integral.

3. Different choices for the placement and weighting of the functional evaluations can give dramatic differences in the rate the numerical methods converge to the exact integral.

3.1.3 Choosing *k* Points in the Brillouin Zone

Because integrals such as Eq. (3.6) take up so much of the computational effort of DFT calculations, it is not surprising that the problem of efficiently evaluating these integrals has been studied very carefully. The solution that is used most widely was developed by Monkhorst and Pack in 1976. Most DFT packages offer the option of choosing *k* points based on this method. To use this method, all that is needed is to specify how many *k* points are to be used in each direction in reciprocal space. For calculations with supercells that have the same length along each lattice vector, and therefore the same length along each reciprocal lattice vector, it is natural to use the same number of *k* points in each direction. If M *k* points are used in each direction, it is usual to label the calculations as using $M \times M \times M$ *k* points.

From the general discussion of numerical integration above, it is clear that we should expect that a calculation using $M \times M \times M$ *k* points will give a more accurate result than a calculation with $N \times N \times N$ *k* points if $M > N$. But, in practice, how should we choose how many *k* points to use? Answering this apparently simple question is more subtle than you might first think, as illustrated by the data in Table 3.2. The data in this table are from calculations done in the same way as the fcc Cu calculations described in Section 2.2 at the optimized lattice constant determined from those calculations, that is, a lattice constant of 3.64 Å. Each calculation listed in Table 3.2 used *k* points defined using the Monkhorst–Pack approach with $M \times M \times M$ *k* points. As in Section 2.2, these calculations used a cubic supercell for the fcc material containing four distinct atoms rather than the primitive cell. Selected results from this table are shown graphically in Fig. 3.2.

Look first at the energies listed in Table 3.2 and plotted in Fig. 3.2. When $M > 8$, the total energy is seen to be (almost) independent of the number of *k* points, as we should expect if all our calculations in *k* space are numerically well converged. More specifically, the variation in the energy as M is varied in this range is less than 0.003 eV. For smaller numbers of *k* points, however, the energy varies considerably as the number of *k* points is changed—a clear indication that the number of *k* points is insufficient to give a well-converged result.

TABLE 3.2   **Results from Computing the Total Energy of fcc Cu with**
$M \times M \times M$ **k Points Generated Using the Monkhorst–Pack Method**

M	E/atom (eV)	No. of k Points in IBZ	τ_M/τ_1
1	-1.8061	1	1.0
2	-3.0997	1	1.1
3	-3.6352	4	2.3
4	-3.7054	4	2.6
5	-3.7301	10	5.2
6	-3.7541	10	6.0
7	-3.7676	20	10.4
8	-3.7671	20	11.2
9	-3.7680	35	16.9
10	-3.7676	35	17.1
11	-3.7662	56	31.2
12	-3.7665	56	28.5
13	-3.7661	84	40.0
14	-3.7659	84	39.7

The last column in Table 3.2 lists the computational time taken for the total energy calculations, normalized by the result for $M = 1$. Note that getting what we just defined to be a converged result takes at least 20 times longer than a calculation involving just one k point. An initially curious feature of these results is that if M is an odd number, then the amount of time taken for the calculations with either M or $(M + 1)$ was close to the same. This occurs because the calculations take full advantage of the many symmetries that exist in a

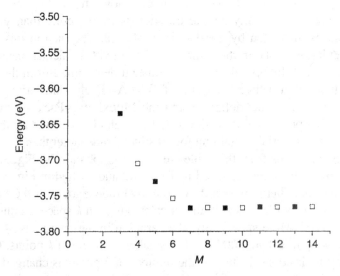

Figure 3.2   Total energies (per atom) for bulk Cu calculated as described in Table 3.2 as a function of M for calculations using $M \times M \times M$ k points. Results with even (odd) M are shown with unfilled (filled) symbols.

[handwritten annotations: even # of k-points get slightly better convergence ↓ explicit symmetry more than odd #s.]

perfect fcc solid. These symmetries mean tha[t] do not need to be evaluated using the entire E uated in a reduced portion of the zone that approximation to fill the entire BZ using sy: ....region in k space is called the *irreducible Brillouin zone* (IBZ). For very symmetric materials such as the perfect fcc crystal, using just the IBZ greatly reduces the numerical effort required to perform integrals in k space. For example, for the $10 \times 10 \times 10$ Monkhorst–Pack sampling of the BZ, only 35 distinct points in k space lie within the IBZ for our current example (compared to the 1000 that would be used if no symmetry at all was used in the calculation).

Table 3.2 lists the number of k points in the IBZ for each calculation. Comparing these with the timings also listed in the table explains why pairs of calculations with odd and even values of M took the same time—they have the same number of distinct points to examine in the IBZ. This occurs because in the Monkhorst–Pack approach using an odd value of M includes some k points that lie on the boundaries of the IBZ (e.g., at the Γ point) while even values of M only give k points inside the IBZ. An implication of this observation is that when small numbers of k points are used, we can often expect slightly better convergence with the same amount of computational effort by using even values of M than with odd values of M.[†] Of course, it is always best to have demonstrated that your calculations are well converged in terms of k points. If you have done this, the difference between even and odd numbers of k points is of limited importance.

To show how helpful symmetry is in reducing the work required for a DFT calculation, we have repeated some of the calculations from Table 3.2 for a four-atom supercell in which each atom was given a slight displacement away from its fcc lattice position. These displacements were not large—they only changed the nearest-neighbor spacings between atoms by ± 0.09 Å, but they removed all symmetry from the system. In this case, the number of k points in the IBZ is $M^3/2$. The results from these calculations are listed in Table 3.3. This table also lists ΔE, the energy difference between the symmetric and nonsymmetric calculations.

Although the total computational time in the examples in Table 3.3 is most closely related to the number of k points in the IBZ, the convergence of the calculations in k space is related to the density of k points in the full BZ. If we compare the entries in Table 3.3 with Fig. 3.2, we see that the calculation for the nonsymmetric system with $M = 8$ is the only entry in the table that might be considered moderately well converged. Further calculations with larger numbers of k points would be desirable if a highly converged energy for the nonsymmetric system was needed.

[†]In some situations such as examining electronic structure, it can be important to include a k point at the Γ point.

TABLE 3.3   Results from Computing the Total Energy of the Variant of fcc Cu with Broken Symmetry[a]

M	E/atom (eV)	ΔE/atom (eV)	No. of k Points in IBZ	τ_M/τ_1
1	-1.8148	-0.009	1	1.0
2	-3.0900	0.010	4	2.1
3	-3.6272	0.008	14	5.6
4	-3.6969	0.009	32	12.3
5	-3.7210	0.009	63	21.9
6	-3.7446	0.010	108	40.1
7	-3.7577	0.010	172	57.5
8	-3.7569	0.010	256	86.8

[a]Described in the text with $M \times M \times M$ k points generated using the Monkhorst–Pack method.

You may have noticed that in Table 3.3 the column of energy differences, ΔE, appears to converge more rapidly with the number of k points than the total energy E. This is useful because the energy difference between the two configurations is considerably more physically interesting than their absolute energies. How does this happen? For any particular set of k points, there is some systematic difference between our numerically evaluated integrals for a particular atomic configuration and the true values of the same integrals. If we compare two configurations of atoms that are structurally similar, then it is reasonable to expect that this systematic numerical error is also similar. This means that the energy difference between the two states can be expected to cancel out at least a portion of this systematic error, leading to calculated energy differences that are more accurate than the total energies from which they are derived. It is important to appreciate that this heuristic argument relies on the two configurations of atoms being "similar enough" that the systematic error from using a finite number of k points for each system is similar. For the example in Table 3.3, we deliberately chose two configurations that differ only by small perturbations of atom positions in the supercell, so it is reasonable that this argument applies. It would be far less reasonable, however, to expect this argument to apply if we were comparing two significantly different crystal structures for a material.

There are many examples where it is useful to use supercells that do not have the same length along each lattice vector. As a somewhat artificial example, imagine we wanted to perform our calculations for bulk Cu using a supercell that had lattice vectors

$$\mathbf{a}_1 = a(1,0,0), \quad \mathbf{a}_2 = a(0,1,0), \quad \mathbf{a}_3 = a(0,0,4). \tag{3.9}$$

calculations that have similar densities of κ points in reciprocal space will have similar levels of convergence. This rule of thumb suggests that using $8 \times 8 \times 2$ k points would give reasonable convergence, where the three numbers refer to the number of k points along the reciprocal lattice vectors $\mathbf{b}_1$, $\mathbf{b}_2$, and $\mathbf{b}_3$, respectively. You should check from the definition of the reciprocal lattice vectors that this choice for the k points will define a set of k points with equal density in each direction in reciprocal space.

3.1.4 Metals—Special Cases in k Space

When we introduced the idea of numerical integration, we looked at the integral $\int_{-1}^{1} (\pi x / 2) \sin(\pi x) \, dx$ as an example. A feature of this problem that we did not comment on before is that it is an integral of a continuous function. This is a useful mathematical property in terms of developing numerical methods that converge rapidly to the exact result of the integral, but it is not a property that is always available in the k space integrals in DFT calculations. An especially important example of this observation is for metals. One useful definition of a metal is that in a metal the Brillouin zone can be divided into regions that are occupied and unoccupied by electrons. The surface in k space that separates these two regions is called the *Fermi surface*. From the point of view of calculating integrals in k space, this is a significant complication because the functions that are integrated change discontinuously from nonzero values to zero at the Fermi surface. If no special efforts are made in calculating these integrals, very large numbers of k points are needed to get well-converged results.

Metals are a rather important subset of all materials, so useful algorithms to improve the slow convergence just mentioned have been developed. We will describe the two best-known methods. The first is called the tetrahedron method. The idea behind this method is to use the discrete set of k points to define a set of tetrahedra that fill reciprocal space and to define the function being integrated at every point in a tetrahedron using interpolation. At the simplest level, linear interpolation can be used within each tetrahedron. Once this interpolation has been completed, the function to be integrated has a simple form at all positions in k space and the integral can then be evaluated using the entire space, not just the original discrete points. Blöchl developed a version of this method that includes interpolation that goes beyond linear interpolation; this is now the most widely used tetrahedron method (see Further Reading).

A different approach to the discontinuous integrals that appear for metals are the smearing methods. The idea of these methods is to force the function

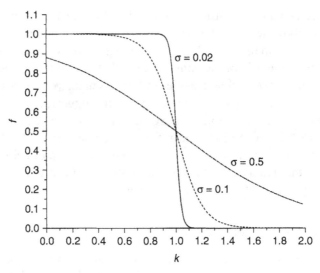

Figure 3.3   Fermi–Dirac function [Eq. (3.10)] with $k_0 = 1$ and several values of σ.

being integrated to be continuous by "smearing" out the discontinuity. An example of a smearing function is the Fermi–Dirac function:

$$f\left(\frac{k - k_0}{\sigma}\right) = \left[\exp\left(\frac{k - k_0}{\sigma}\right) + 1\right]^{-1}. \qquad (3.10)$$

Figure 3.3 shows the shape of this function for several values of σ. It can be seen from the figure that as $\sigma \to 0$, the function approaches a step function that changes discontinuously from 1 to 0 at $k = k_0$. The idea of using a smearing method to evaluate integrals is to replace any step functions with smooth functions such as the Fermi–Dirac function since this defines a continuous function that can be integrated using standard methods. Ideally, the result of the calculation should be obtained using some method that extrapolates the final result to the limit where the smearing is eliminated (i.e., $\sigma \to 0$ for the Fermi–Dirac function).

One widely used smearing method was developed by Methfessel and Paxton. Their method uses expressions for the smearing functions that are more complicated than the simple example above but are still characterized by a single parameter, σ (see Further Reading).

3.1.5 Summary of k Space

Because making good choices about how reciprocal space is handled in DFT calculations is so crucial to performing meaningful calculations, you should read the following summary carefully and come back to it when you start doing calculations on your own. The key ideas related to getting well-converged results in k space include:

1. Before pursuing a large series of DFT calculations for a system of interest, numerical data exploring the convergence of the calculations with respect to the number of k points should be obtained.
2. The number of k points used in any calculation should be reported since not doing so makes reproduction of the result difficult.
3. Increasing the volume of a supercell reduces the number of k points needed to achieve convergence because volume increases in real space correspond to volume decreases in reciprocal space.
4. If calculations involving supercells with different volumes are to be compared, choosing k points so that the density of k points in reciprocal space is comparable for the different supercells is a useful way to have comparable levels of convergence in k space.
5. Understanding how symmetry is used to reduce the number of k points for which calculations are actually performed can help in understanding how long individual calculations will take. But overall convergence is determined by the density of k points in the full Brillouin zone, not just the number of k points in the irreducible Brillouin zone.
6. Appropriate methods must be used to accurately treat k space for metals.

3.2 ENERGY CUTOFFS

Our lengthy discussion of k space began with Bloch's theorem, which tells us that solutions of the Schrödinger equation for a supercell have the form

$$\phi_{\mathbf{k}}(\mathbf{r}) = \exp(i\mathbf{k} \cdot \mathbf{r}) u_{\mathbf{k}}(\mathbf{r}), \tag{3.11}$$

where $u_{\mathbf{k}}(\mathbf{r})$ is periodic in space with the same periodicity as the supercell. It is now time to look at this part of the problem more carefully. The periodicity of $u_{\mathbf{k}}(\mathbf{r})$ means that it can be expanded in terms of a special set of plane waves:

$$u_{\mathbf{k}}(\mathbf{r}) = \sum_{\mathbf{G}} c_{\mathbf{G}} \exp[i\mathbf{G} \cdot \mathbf{r}], \tag{3.12}$$

where the summation is over all vectors defined by $\mathbf{G} = m_1 \mathbf{b}_1 + m_2 \mathbf{b}_2 + m_3 \mathbf{b}_3$ with integer values for m_i. These set of vectors defined by $\mathbf{G}$ in reciprocal space are defined so that for any real space lattice vector $\mathbf{a}_i$, $\mathbf{G} \cdot \mathbf{a}_i = 2\pi m_i$.

(You can check this last statement by using the definition of the reciprocal lattice vectors.)

Combining the two equations above gives

$$\phi_{\mathbf{k}}(\mathbf{r}) = \sum_{\mathbf{G}} c_{\mathbf{k}+\mathbf{G}} \exp[i(\mathbf{k} + \mathbf{G})\mathbf{r}]. \tag{3.13}$$

According to this expression, evaluating the solution at even a single point in k space involves a summation over an infinite number of possible values of $\mathbf{G}$. This does not sound too promising for practical calculations! Fortunately, the functions appearing in Eq. (3.13) have a simple interpretation as solutions of the Schrödinger equation: they are solutions with kinetic energy

$$E = \frac{\hbar^2}{2m} |\mathbf{k} + \mathbf{G}|^2.$$

It is reasonable to expect that the solutions with lower energies are more physically important than solutions with very high energies. As a result, it is usual to truncate the infinite sum above to include only solutions with kinetic energies less than some value:

$$E_{\text{cut}} = \frac{\hbar^2}{2m} G_{\text{cut}}^2.$$

The infinite sum then reduces to

$$\phi_{\mathbf{k}}(\mathbf{r}) = \sum_{|\mathbf{G}+\mathbf{k}|<G_{\text{cut}}} c_{\mathbf{k}+\mathbf{G}} \exp[i(\mathbf{k} + \mathbf{G})\mathbf{r}]. \tag{3.14}$$

This expression includes slightly different numbers of terms for different values of $\bar{\mathbf{k}}$.

The discussion above has introduced one more parameter that must be defined whenever a DFT calculation is performed—the cutoff energy, E_{cut}. In many ways, this parameter is easier to define than the k points, as most packages will apply sensible default settings if no other information is supplied by the user. Just as with the k points, it is good practice to report the cutoff energy used in your calculations to allow people to reproduce your results easily. Figure 3.4 shows as an example of the convergence of the energy of fcc Cu as the energy cutoff in our calculations was varied.

There is one common situation where it is unwise to blindly accept default values for the cutoff energy. In most cases, a default cutoff energy is

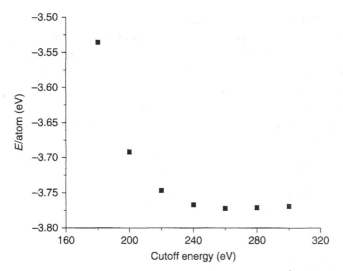

Figure 3.4 Energy per atom of fcc Cu with a lattice constant of 3.64 Å using $12 \times 12 \times 12$ k points as a function of the energy cutoff, plotted using a similar energy scale to Fig. 3.2.

assigned for each element and the largest cutoff energy for any of the atoms in the supercell is assigned as the overall cutoff energy. Suppose we used this approach, for example, while attempting to calculate the energy change in forming an ordered body-centered cubic (bcc) copper-palladium alloy from bulk palladium and copper, fcc $Pd_{(s)}$ + fcc $Cu_{(s)} \rightarrow$ bcc $CuPd_{(s)}$. We can calculate this energy change by finding well-converged energies for each of the solids separately. (Exercise 7 at the end of the chapter asks you to do exactly this.) For the purposes of this example, we will assume that the energy cutoffs for Pd and Cu satisfy $E_{Pd} > E_{Cu}$. If we used the default cutoff energies, we would calculate the energies of fcc Pd and bcc CuPd using an energy cutoff of E_{Pd}, but the energy of fcc Cu would be calculated using an energy cutoff of E_{Cu}. This approach introduces a systematic error into the calculated energy difference that can easily be removed by performing the fcc Cu calculation using the higher cutoff energy, E_{Pd}. The key point to remember here is *whenever DFT calculations for multiple systems are compared to calculate energy differences, the same energy cutoff should be used in all calculations.*

3.2.1 Pseudopotentials

The discussion above points to the fact that large energy cutoffs must be used to include plane waves that oscillate on short length scales in real space. This is problematic because the tightly bound core electrons in atoms are associated with wave functions with exactly this kind of oscillation. From a physical point of view, however, core electrons are not especially important in defining chemical bonding and other physical characteristics of materials; these properties are dominated by the less tightly bound valence electrons. From the earliest developments of plane-wave methods, it was clear that there could be great advantages in calculations that approximated the properties of core electrons in a way that could reduce the number of plane waves necessary in a calculation.

The most important approach to reducing the computational burden due to core electrons is to use pseudopotentials. Conceptually, a pseudopotential replaces the electron density from a chosen set of core electrons with a smoothed density chosen to match various important physical and mathematical properties of the true ion core. The properties of the core electrons are then fixed in this approximate fashion in all subsequent calculations; this is the frozen core approximation. Calculations that do not include a frozen core are called all-electron calculations, and they are used much less widely than frozen core methods. Ideally, a pseudopotential is developed by considering an isolated atom of one element, but the resulting pseudopotential can then be used reliably for calculations that place this atom in any chemical environment without further adjustment of the pseudopotential. This desirable property is referred to as the transferability of the pseudopotential. Current DFT codes typically provide a library of pseudopotentials that includes an entry for each (or at least most) elements in the periodic table.

The details of a particular pseudopotential define a minimum energy cutoff that should be used in calculations including atoms associated with that pseudopotential. Pseudopotentials requiring high cutoff energies are said to be hard, while more computationally efficient pseudopotentials with low cutoff energies are soft. The most widely used method of defining pseudopotentials is based on work by Vanderbilt; these are the ultrasoft pseudopotentials (USPPs). As their name suggests, these pseudopotentials require substantially lower cutoff energies than alternative approaches.

One disadvantage of using USPPs is that the construction of the pseudopotential for each atom requires a number of empirical parameters to be specified. Current DFT codes typically only include USPPs that have been carefully developed and tested, but they do in some cases include multiple USPPs with varying degrees of softness for some elements. Another frozen core approach that avoids some of the disadvantages of USPPs is the projector augmented-wave (PAW) method originally introduced by Blöchl and later adapted for plane-wave calculations by Kresse and Joubert. Kresse and Joubert performed an extensive comparison of USPP, PAW, and all-electron calculations for small molecules and extended solids.[1] Their work shows that well-constructed USPPs and the PAW method give results that are essentially identical in many cases and, just as importantly, these results are in good agreement with all-electron calculations. In materials with strong magnetic moments or with atoms that have large differences in electronegativity, the PAW approach gives more reliable results than USPPs.

3.3 NUMERICAL OPTIMIZATION

All of our work on numerical convergence in this chapter has so far focused on finding numerically precise solutions for the electron density and total energy of a configuration of atoms within DFT. If you think back to the examples in Chapter 2, you will soon remember that we compared a series of total energy calculations to predict the most stable crystal structure and lattice constant for a simple material, bulk Cu. One of these examples, Cu in the hcp crystal structure, illustrated the idea that when a crystal structure has more than one degree of freedom, finding the minimum energy structure by systematically varying the degrees of freedom is, at best, tedious.

To make practical use of our ability to perform numerically converged DFT calculations, we also need methods that can help us effectively cope with situations where we want to search through a problem with many degrees of freedom. This is the topic of numerical optimization. Just like understanding reciprocal space, understanding the central concepts of optimization is vital to using DFT effectively. In the remainder of this chapter, we introduce these ideas.

3.3.1 Optimization in One Dimension

We begin with a topic that is crucial to performing DFT calculations but has ramifications in many other areas as well. A simple starting point is the following mathematical problem: find a local minimum of $f(x)$, where $f(x)$ is some function of x that is uniquely defined at each x. We will assume that $f(x)$ is a "smooth" function, meaning that its derivatives exist and are continuous. You should also think of an example where $f(x)$ is so complicated that you cannot possibly solve the problem algebraically. In other words, we have to use a computer to find a solution to the problem numerically. Note that we are asking for "a" local minimum, not "all" local minima or the lowest possible value of the function. This means we are trying to solve a local optimization problem, not a global optimization problem.

It is hopefully not hard to see why this mathematical problem (or closely related problems) are important in DFT calculations. The problem of finding the lattice constant for an fcc metal that we looked at in Chapter 2 can be cast as a minimization problem. If we do not know the precise arrangement of N atoms within a supercell, then the total energy of the supercell can be written as $E(\mathbf{x})$, where $\mathbf{x}$ is a $3N$-dimensional vector defining the atomic positions. Finding the physically preferred arrangement of atoms is equivalent to minimizing this energy function.[*]

For now we will consider the original one-dimensional problem stated above: Find a local minimum of $f(x)$. If we can find a value of x where $f'(x) = df/dx = 0$, then this point defines either a local maximum or a minimum of the function. Since it is easy to distinguish a maximum from a minimum (either by looking at the second derivative or more simply by evaluating the function at some nearby points), we can redefine our problem as: find a value of x where $f'(x) = 0$. We will look at two numerical methods to solve this problem. (There are many other possible methods, but these two will illustrate some of the key concepts that show up in essentially all possible methods.)

The first approach is the *bisection method*. This method works in more or less the way you find a street address when driving in an unfamiliar neighborhood on a dark rainy night. In that case, you try and establish that the address you want is somewhere between two points on the street, then you look more carefully between those two points. To be more specific, we begin by finding two points, $x_1 = a$ and $x_2 = b$ so that the sign of $f'(x)$ at the two points is different: $f'(x_1)f'(x_2) < 0$. As long as $f'(x)$ is smooth, we then know that there is some x between a and b where $f'(x) = 0$. We do not know where this point is, so we look right in the middle by evaluating $f'(x^*)$ at $x^* = (x_1 + x_2)/2$. We can now see one of two things. If $f'(x_1)f'(x^*) < 0$, then we have established that $f'(x) = 0$ for some x between x_1 and x^*. Alternatively, if $f'(x^*)f'(x_2) < 0$, then $f'(x) = 0$ for some x between x^* and x_2. You should convince yourself by drawing a picture that if there is only one place where $f'(x) = 0$ in the interval $[a,b]$ that exactly one of these two possibilities can occur (not both of them or neither of them). In either case, we have reduced the size of the interval in which we are searching by a factor of 2. We can now repeat the whole procedure for this new interval. Each time the calculation is repeated, a new value of x^* is generated, and if we continue repeating the calculation long enough, these values of x^* will get closer and closer to the situation we are looking for where $f'(x) = 0$.

As a simple example, we can apply the bisection method to find a minimum of $f(x) = e^{-x}\cos x$. This function has a minimum at $x = 2.356194490\ldots$. If we use the bisection method starting with $x_1 = 1.8$ and $x_2 = 3$, then we generate the following series of approximations to the solution: $x^* = 2.4$, 2.1, 2.25, 2.325, 2.3625, 2.34375, $\ldots$. These are getting closer to the actual solution as we continue to repeat the calculation, although after applying

---

[*]As we will see in Section 3.4, minimization is also crucial in other aspects of DFT calculations.

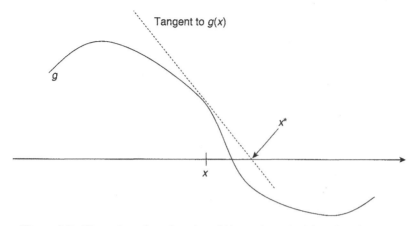

Figure 3.5   Illustration of one iteration of Newton's method for a function $g(x)$.

the algorithm six times we still only have a solution that is accurate to two significant digits.

An alternative approach to our problem is *Newton's method*. The idea behind this method is illustrated in Fig. 3.5. If we define $g(x) = f'(x)$, then from a Taylor expansion $g(x + h) \cong g(x) + hg'(x)$. This expression neglects terms with higher orders of h, so it is accurate for "small" values of h. If we have evaluated our function at some position where $g(x) \neq 0$, our approximate equation suggests that a good estimate for a place where the function *is* zero is to define $x^* = x + h = x - g(x)/g'(x)$. Because the expression from the Taylor expansion is only approximate, x^* does not exactly define a place where $g(x^*) = 0$. Just as we did with the bisection method, we now have to repeat our calculation. For Newton's method, we repeat the calculation starting with the estimate from the previous step.

Applying Newton's method to finding a minimum of $f(x) = e^{-x} \cos x$ starting with $x = 1.8$ generates the following series of approximations to the solution: $x^* = 1.8$, 2.183348, 2.331986, 2.355627, 2.356194, ... (showing seven significant figures for each iterate). Just as with the bisection method, this series of solutions is converging to the actual value of x that defines a minimum of the function. A striking difference between the bisection method and Newton's method is how rapidly the solutions converge. A useful way to characterize the rate of convergence is to define $\varepsilon_i = |x_i - x|$, where x_i is the approximation to the true solution x after applying the numerical method i times. Figure 3.6 plots the result for the two algorithms. It is clear from this figure that Newton's method is enormously better than the bisection method in terms of how rapidly it allows us to find the solution.

There is one crucial aspect of these numerical methods that we have not yet discussed. Both methods generate a series of approximations that in the

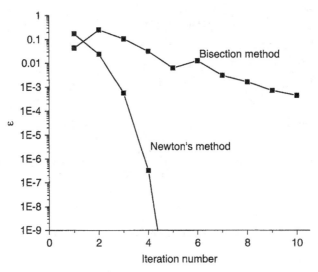

Figure 3.6 Convergence analysis for the bisection method and Newton's method calculations described in the text.

long run get closer and closer to a solution. How do we know when to stop? In a real application, we do not know the exact solution, so we cannot use $\varepsilon_i = |x_i - x|$ as our measure of how accurate our current estimate is. Instead, we have to define a stopping criteria that depends only on the series of approximations we have calculated. A typical choice is to continue iterating until the difference between successive iterates is smaller than some tolerance, that is, $|x_{i+1} - x_i| < \delta$. If we chose $\delta = 0.01$ for the examples above, we would stop after 6 iterations of the bisection method, concluding that $x \cong 2.34375$, or after 3 iterations of Newton's method, concluding that $x \cong 2.355628$. If we chose a smaller tolerance, $\delta = 0.001$, we would stop after 11 (4) iterations of the bisection method (Newton's method) with the approximation that $x \cong 2.3554688$ ($x \cong 2.3561942$).

The examples above demonstrate several general properties of numerical optimization that are extremely important to appreciate. They include:

1. The algorithms are iterative, so they do not provide an exact solution; instead, they provide a series of approximations to the exact solution.
2. An initial estimate for the solution must be provided to use the algorithms. The algorithms provide no guidance on how to choose this initial estimate.
3. The number of iterations performed is controlled by a tolerance parameter that estimates how close the current solution is to the exact solution.

4. Repeating any algorithm with a different tolerance parameter or a different initial estimate for the solution will generate multiple final approximate solutions that are (typically) similar *but are not identical.*

5. The rate at which different algorithms converge to a solution can vary greatly, so choosing an appropriate algorithm can greatly reduce the number of iterations needed.

There are other general properties of these methods that have not been directly illustrated by our examples but are equally important. These include:

6. These methods cannot tell us if there are multiple minima of the function we are considering if we just apply the method once. Applying a method multiple times with different initial estimates can yield multiple minima, but even in this case the methods do not give enough information to prove that all possible minima have been found. The function we used as an example above was chosen specifically because it has multiple minima. Exercise 1 at the end of the chapter asks you to explore this idea.

7. For most methods, no guarantees can be given that the method will converge to a solution at all for an arbitrary initial estimate. Exercise 2 at the end of the chapter gives a specific example to illustrate this idea.

3.3.2   Optimization in More than One Dimension

We now return to the problem mentioned above of minimizing the total energy of a set of atoms in a supercell, $E(\mathbf{x})$, by varying the positions of the atoms within the supercell. This example will highlight some of the complications that appear when we try and solve multivariable optimization problems that are not present for the one-dimensional situations we have already discussed. If we define $g_i(\mathbf{x}) = \partial E(\mathbf{x})/\partial x_i$, then minimizing $E(\mathbf{x})$ is equivalent to finding a set of positions for which $g_i(\mathbf{x}) = 0$ simultaneously for $i = 1, \ldots, 3N$.

There is no natural way to generalize the one-dimensional bisection method to solve this multidimensional problem. But it is possible to generalize Newton's method to this situation. The one-dimensional Newton method was derived using a Taylor expansion, and the multidimensional problem can be approached in the same way. The result involves a $3N \times 3N$ matrix of derivatives, J, with elements $J_{ij} = \partial g_i/\partial x_j$. Note that the elements of this matrix are the second partial derivatives of the function we are really interested in, $E(\mathbf{x})$. Newton's method defines a series of iterates by

$$\mathbf{x}_{i+1} = \mathbf{x}_i - J^{-1}(\mathbf{x}_i)\mathbf{g}(\mathbf{x}_i). \tag{3.15}$$

This looks fairly similar to the result we derived for the one-dimensional case, but in practice is takes a lot more effort to apply. Consider a situation where we have 10 distinct atoms, so we have 30 distinct variables in our problem. We will have to calculate 55 distinct matrix elements of J during each iteration. The other 45 matrix elements are then known because the matrix is symmetric. We then have to solve a linear algebra problem involving a 30×30 matrix. You should compare this to how much work was involved in taking one step in the one-dimensional case. It is certainly quite feasible to perform all of these calculations for the multidimensional problem computationally. The main point of this discussion is that the amount of numerical effort required to tackle multidimensional optimization grows rather dramatically as the number of variables in the problem is increased.

Another distinction between one-dimensional and multidimensional problems is associated with searching for multiple solutions. If we aim to find all the minima of a one-dimensional function, $f(x)$, in an interval $[a,b]$, we can in principle repeatedly apply Newton's method with many initial estimates in the interval and build up a list of distinct local minima. We cannot rigorously prove that the list we build up in this way is complete, but it is not too difficult for a simple example to do enough calculations to be convinced that no new minima will show up by performing more calculations. The same idea cannot be applied in 30 dimensions simply because of the vast number of possible initial estimates that would have to be considered to fully sample the possible space of solutions. As a result, it is in general extremely difficult to perform optimization in multiple dimensions for problems relevant to DFT calculations in a way that gives any information other than the existence of a local minimum of a function. Algorithms to rigorously find the global minimum value of a function do exist, but they are extraordinarily computationally intensive. To our knowledge, no one has even attempted to use algorithms of this kind in conjunction with DFT calculations.

At this point it may seem as though we can conclude our discussion of optimization methods since we have defined an approach (Newton's method) that will rapidly converge to optimal solutions of multidimensional problems. Unfortunately, Newton's method simply cannot be applied to the DFT problem we set ourselves at the beginning of this section! To apply Newton's method to minimize the total energy of a set of atoms in a supercell, $E(\mathbf{x})$, requires calculating the matrix of second derivatives of the form $\partial^2 E/\partial x_i\, \partial x_j$. Unfortunately, it is very difficult to directly evaluate second derivatives of energy within plane-wave DFT, and most codes do not attempt to perform these calculations. The problem here is not just that Newton's method is numerically inefficient—it just is not practically feasible to evaluate the functions we need to use this method. As a result, we have to look for other approaches to minimize $E(\mathbf{x})$. We will briefly discuss the two numerical methods that are most commonly used for this problem: quasi-Newton and conjugate-gradient methods. In both cases we will only sketch the main concepts. Resources that discuss the details of these methods are listed in the Further Reading section at the end of the chapter.

The essence of quasi-Newton methods is to replace Eq. (3.15) by

$$\mathbf{x}_{i+1} = \mathbf{x}_i - A_i^{-1}(\mathbf{x}_i)\mathbf{g}(\mathbf{x}_i), \tag{3.16}$$

where A_i is a matrix that is defined to approximate the Jacobian matrix. This matrix is also updated iteratively during the calculation and has the form

$$A_i = A_{i-1} + F[\mathbf{x}_i, \mathbf{x}_{i-1}, \mathbf{G}(\mathbf{x}_i), \mathbf{g}(\mathbf{x}_{i-1})]. \tag{3.17}$$

We have referred to quasi-Newton methods rather than the quasi-Newton method because there are multiple definitions that can be used for the function F in this expression. The details of the function F are not central to our discussion, but you should note that this updating procedure now uses information from the current and the previous iterations of the method. This is different from all the methods we have introduced above, which only used information from the current iteration to generate a new iterate. If you think about this a little you will realize that the equations listed above only tell us how to proceed once several iterations of the method have already been made. Describing how to overcome this complication is beyond our scope here, but it does mean than when using a quasi-Newton method, the convergence of the method to a solution should really only be examined after performing a minimum of four or five iterations.

The conjugate-gradient method to minimizing $E(\mathbf{x})$ is based on an idea that is quite different to the Newton-based methods. We will introduce these ideas with a simple example: calculating the minima of a two-dimensional function $E(\mathbf{x}) = 3x^2 + y^2$, where $\mathbf{x} = (x, y)$. Hopefully, you can see right away what the solution to this problem: the function has exactly one minimum and it lies at $\mathbf{x} = (0,0)$. We will try to find this solution iteratively starting from $\mathbf{x}_0 = (1,1)$. Since we are trying to minimize the function, it makes sense to look in a direction that will cause the function to decrease. A basic result from vector calculus is that the function $E(\mathbf{x})$ decreases most rapidly along a direction defined by the negative of the gradient of the function, $-\nabla E(\mathbf{x})$. This suggests we should generate a new estimate by defining

$$\mathbf{x}_1 = \mathbf{x}_0 - \alpha_0 \nabla E(\mathbf{x}_0) = (1 - 6\alpha_0, 1 - 2\alpha_0). \tag{3.18}$$

This expression tells us to look along a particular line in space, but not where on that line we should place our next iterate. Methods of this type are known as steepest descent methods. Ideally, we would like to choose the step length, α_0, so that $E(\mathbf{x}_1)$ is smaller than the result we would get with any other choice for the step length. For our example, a little algebra shows that $E[\mathbf{x}_1(\alpha_0)]$ is minimized if we choose $\alpha_0 = \frac{5}{28}$.

We can now repeat this procedure to generate a second iterate. This time, we want to search along a direction defined by the gradient of $E(\mathbf{x})$ evaluated at $\mathbf{x}_1$: $\nabla E(\mathbf{x}_1) = \left(-\frac{3}{7}, \frac{9}{7}\right)$. The point of defining the details of this equation is to highlight the following observation:

$$\nabla E(\mathbf{x}_0) \cdot \nabla E(\mathbf{x}_1) = 0. \tag{3.19}$$

In our two-dimensional space, these two search directions are perpendicular to one another. Saying this in more general mathematical terms, the two search directions are orthogonal. This is not a coincidence that occurs just for the specific example we have defined; it is a general property of steepest descent methods provided that the line search problem defined by Eq. (3.18) is solved optimally.

We can now qualitatively describe the conjugate-gradient method for minimizing a general function, $E(\mathbf{x})$, where $\mathbf{x}$ is an N-dimensional vector. We begin with an initial estimate, $\mathbf{x}_0$. Our first iterate is chosen to lie along the direction defined by $\mathbf{d}_0 = -\nabla E(\mathbf{x}_0)$, so $\mathbf{x}_1 = \mathbf{x}_0 - \alpha_0 \mathbf{d}_0$. Unlike the simple example above, the problem of choosing the best (or even a good) value of α_0 cannot be solved exactly. We therefore choose the step size by some approximate method that may be as simple as evaluating $E(\mathbf{x}_1)$ for several possible step lengths and selecting the best result.

If we have the optimal step length, then the next search direction will be orthogonal to our current search direction. In general, however, $\nabla E(\mathbf{x}_0)\nabla E(\mathbf{x}_1) \neq 0$ because we cannot generate the optimal step length for each line search. The key idea of the conjugate gradient is to define the new search direction by insisting that it *is* orthogonal to the original search direction. We can do this by defining

$$\mathbf{d}_1 = -\nabla E(\mathbf{x}_1) + \frac{(\nabla E(\mathbf{x}_1) \cdot \mathbf{d}_0)}{(\mathbf{d}_0 \cdot \mathbf{d}_0)} \mathbf{d}_0. \tag{3.20}$$

This definition uses $\nabla E(\mathbf{x}_1)$ as an estimate for the new search direction but removes the portion of this vector that can be projected along the original search direction, $\mathbf{d}_0$. This process is called orthogonalization, and the resulting search direction is a conjugate direction to the original direction (hence the name of the overall method). We now generate a second iterate by looking at various step lengths within $\mathbf{x}_2 = \mathbf{x}_1 - \alpha_1 \mathbf{d}_1$.

In generating a third iterate for the conjugate-gradient method, we now estimate the search direction by $-\nabla E(\mathbf{x}_2)$ but insist that the search direction is orthogonal to both $\mathbf{d}_0$ and $\mathbf{d}_1$. This idea is then repeated for subsequent iterations.

We cannot continue this process indefinitely because in an N-dimensional space we can only make a vector orthogonal to at most $(N-1)$ other vectors. So to make this a well-defined algorithm, we have to restart the process of defining search directions after some number of iterations less than N.

The conjugate-gradient method is a powerful and robust algorithm. Using it in a practical problem involves making choices such as how to select step lengths during each iteration and how often to restart the process of orthogonalizing search directions. Like the quasi-Newton method, it can be used to minimize functions by evaluating only the function and its first derivatives.

3.3.3   What Do I Really Need to Know about Optimization?

We have covered a lot of ground about numerical optimization methods, and, if you are unfamiliar with these methods, then it is important to distill all these details into a few key observations. This is not just a mathematical tangent because performing numerical optimization efficiently is central to getting DFT calculations to run effectively. We highly recommend that you go back and reread the seven-point summary of properties of numerical optimization methods at the end of Section 3.3.1. You might think of these (with apologies to Stephen Covey[2]) as the "seven habits of effective optimizers." These points apply to both one-dimensional and multidimensional optimization. We further recommend that as you start to actually perform DFT calculations that you actively look for the hallmarks of each of these seven ideas.

3.4   DFT TOTAL ENERGIES—AN ITERATIVE OPTIMIZATION PROBLEM

The most basic type of DFT calculation is to compute the total energy of a set of atoms at prescribed positions in space. We showed results from many calculations of this type in Chapter 2 but have not said anything about how they actually are performed. The aim of this section is to show that this kind of calculation is in many respects just like the optimization problems we discussed above.

As we discussed in Chapter 1, the main aim of a DFT calculation is to find the electron density that corresponds to the ground-state configuration of the system, $\rho(\mathbf{r})$. The electron density is defined in terms of the solutions to the Kohn–Sham equations, $\psi_j(\mathbf{r})$, by

$$\rho(\mathbf{r}) = \sum_j \psi_j(\mathbf{r})\psi_j^*(\mathbf{r}). \tag{3.21}$$

The Kohn–Sham equations are

$$-\frac{\hbar^2}{2m}\nabla^2\psi_j(\mathbf{r}) + V_{\text{eff}}(\mathbf{r})\psi_j(\mathbf{r}) = \varepsilon_j\psi_j(\mathbf{r}), \qquad (3.22)$$

where $V_{\text{eff}}(\mathbf{r})$ is the effective potential. The fact that makes these equations awkward to solve directly is that the effective potential is itself a complicated function of $\rho(\mathbf{r})$.

Our introduction to numerical optimization suggests a useful general strategy for solving the problem just posed, namely, attempt to solve the problem iteratively. We begin by estimating the overall electron density, then use this trial density to define the effective potential. The Kohn–Sham equations with this effective potential are then solved numerically with this effective potential, defining a new electron density. If the new electron density and the old electron density do not match, then we have not solved the overall problem. The old and new electron densities are then combined in some way to give a new trial electron density. This new trial density is then used to define a new effective potential from which an updated electron density is found, and so on. If successful, this iterative process will lead to a self-consistent solution. This description has glossed over all kinds of numerical details that were absolutely vital in making modern DFT codes numerically efficient.

The similarity of this iterative process to the more general optimization problems we talked about above suggest that our "seven habits of effective optimizers" might give us some useful ideas about how to calculate energies with DFT. In particular, we need to think about how to start our iterative process. In the absence of other information, the electron density can be initially approximated by superimposing the electron densities appropriate for each atom in its isolated state. This is typically the default initialization used by most DFT packages. But we will reach a self-consistent solution much more quickly (i.e., in fewer iterations) if a better initial approximation is available. So if we have previously calculated the electron density for a situation very similar to our current atomic configuration, that electron density may be a useful initial approximation. For this reason, it can sometimes be helpful to store the electron density and related information from a large calculation for use in starting a subsequent similar calculation.

We also need to think about how to stop our iterative calculations. It is not necessarily convenient to directly compare two solutions for the electron density and determine how similar they are, even though this is the most direct test for whether we have found a self-consistent solution. A method that is easier to interpret is to calculate the energy corresponding to the electron density after each iteration. This is, after all, the quantity we are ultimately interested in finding. If our iterations are converging, then the difference in energy between consecutive iterates will approach zero. This suggests that the iterations can be stopped once the magnitude of the energy difference between iterates falls below an appropriately chosen tolerance parameter. Most DFT packages define a sensible default value for this parameter, but for high-precision calculations it may be desirable to use lower values. A general feature of the iterative algorithms used in calculations of this sort is that they converge rapidly once a good approximation for the electron density is found. This means that reducing the tolerance parameter that defines the end of a self-consistent calculation by even several orders of magnitude often leads to only a handful of additional iterative steps.

Before moving on from this section, it would be a good idea to understand within the DFT package that is available to you (i) what algorithms are used for solving the self-consistent Kohn–Sham problem and (ii) how you can verify from the output of a calculation that a self-consistent solution was reached.

3.5 GEOMETRY OPTIMIZATION

3.5.1 Internal Degrees of Freedom

We have so far only described DFT calculations in which the position of every atom in the supercell is fixed. This is fine if we only want to predict the lattice constants of simple solids, but other than that it can only give us a limited view of the world! Let us imagine, for example, that we are interested in a set of reactions involving nitrogen. One thing we would certainly like to know would be the energy of a molecule of gas-phase N_2 since this will probably be needed for describing any overall reaction that begins with gaseous N_2. To calculate this energy, we need to find the geometry of N_2 that minimizes the molecule's total energy. Because this is such a simple molecule, this task means that we have to determine a single bond length. How can we do this using DFT calculations based on periodic supercells?

To mimic a gas-phase molecule, we need to build a supercell that is mainly empty space. One simple way to do this is to define a cubic supercell with side length L angstroms and place two N atoms in the supercell with fractional coordinates $(0,0,0)$ and $(+d/L,0,0)$. So long as L is considerably longer than d, this supercell represents an isolated N_2 molecule with bond length d.* We can now find the DFT-optimized bond length for N_2 by using either the quasi-Newton or conjugate-gradient methods defined above to minimize the total energy of our supercell, fixing the size and shape of the supercell but allowing the fractional coordinates of the two atoms to vary. To do this, we must define the stopping criterion that will be used to decide whether these iterative schemes have converged to a minimum. Because we are

---

*As long as the supercell is large enough, the influence of the periodic images of the molecule on the total energy will be small, particularly for a molecule with no dipole moment like N_2.

searching for a configuration where the forces on both atoms are zero, we continue our iterative calculations until the magnitude of the force on both atoms is less than $0.01 \text{ eV}/\text{Å}$. In this example, the magnitude of the forces on the two atoms are always identical by symmetry, although, of course, the forces on the two atoms point in opposite direction. Why is this force criterion reasonable? If we change an atom's position by a small amount, Δr, then the change in total energy due to this change can be estimated by $|\Delta E| \cong |\mathbf{F}||\Delta r|$, where $|\mathbf{F}|$ is the force on the atom. If the forces on all atoms are less than $0.01 \text{ eV}/\text{Å}$, then moving any individual atom by 0.1 Å, a relatively significant distance in terms of chemical bond lengths, will change the total energy by less than 0.001 eV, a small amount of energy. This argument only tells us that the order of magnitude of the force in the stopping criterion is reasonable.

The second choice we must make is how far apart to place the atoms initially. It may seem like this choice is not too important since we are going to minimize the energy, and there is presumably only one bond length that actually defines this energy minimum. Unfortunately, choosing the initial estimate for the bond length can make a large impact on whether the calculation is successful or not. As an example, we will consider two calculations that used a conjugate-gradient optimization method that differed only in the initial distance between the atoms. When we initially placed the atoms 1 Å apart, our calculations proceeded for 11 conjugate-gradient iterations, after which the forces satisfied the criterion above and the bond length was 1.12 Å. The experimentally observed N_2 bond length is 1.10 Å. As with the lattice parameters of bulk solids that we examined in Chapter 2, the DFT result is not exact, but the difference between the DFT optimized geometry and the experimental result is small.

In a second calculation, we initially placed the atoms 0.7 Å apart. In this case, after 25 conjugate-gradient steps then two atoms are 2.12 Å apart and, even worse, the algorithm has not converged to an energy minimum. What has happened? Let us think about our initial state with the two atoms 0.7 Å apart. This corresponds physically to a N_2 molecule with its bond enormously compressed relative to its normal value. This means that there is an enormous repulsive force pushing the two atoms away from each other. This matters because our optimization methods involve estimating how rapidly the total energy changes based on derivatives of the energy evaluated at a single location. The numerical optimization method recognizes that the two atoms want to move away from each other and obliges by taking an initial step that separates the two atoms by a large distance.

Unfortunately, the two atoms are now in positions that in some sense have given an even worse approximation to the minimum energy state of the molecule than our initial estimate and the calculation is unable to recover and find the true minimum. In short, our calculation has failed miserably because we used an initial geometry that was chemically implausible. The details of how this type of failure will unfold are dependent on the details of the optimization algorithm, but this is something that can and will happen with almost any optimization method if you use poor initial estimates for the geometries of the atoms in which you are interested. The critical lesson here is that expending effort to create good initial estimates for atomic coordinates will greatly increase the speed of your DFT calculations and in many cases make a decisive difference in whether your calculations can even converge to an energy minimum.

As a second example of optimizing the positions of atoms within a supercell, we will optimize the geometry of a molecule of CO_2. If we again use a cubic supercell with side length L angstroms, we can create a CO_2 molecule by placing a C atom at fractional coordinates $(0,0,0)$ and O atoms at $(+d/L,0,0)$ and $(-d/L,0,0)$. Optimizing the energy of the molecule from this initial state with $d = 1.3$ Å and the same stopping criterion that we used for the N_2 calculations gives us an optimized C–O bond length of 1.17 Å and an O–C–O bond angle of $180°$.

Our CO_2 results seem quite reasonable, but can we trust this result? Remember that we defined our stopping criterion by the magnitude of the force on the atoms. Let us examine the force on one of the O atoms for the configuration we used in our calculations. If we write this force as $\mathbf{f} = (f_x, f_y, f_z)$, then *by symmetry*, $f_y = f_z = 0$, regardless of what value of d we choose. This means that as the geometry of the molecule is iteratively updated during energy minimization, the C–O bond lengths will be varied but the O–C–O bond angle will remain fixed at $180°$, the value we defined in our initial configuration. Saying this another way, the bond angle in our calculation is not really a converged result, it is an inevitable result of the symmetry we imposed in our original estimate for the molecule's structure.

A more reliable approach to optimizing the geometry of this molecule is to choose an initial bond angle that does not make components of the forces vanish by symmetry alone. We can easily do this by starting from a configuration with a C atom at fractional coordinates $(0,0,0)$ and O atoms at $(+a/L, b/L,0)$ and $(-a/L, b/L,0)$. Minimizing the energy of the molecule starting from this structure with $a = 1.2$ Å and $b = 0.1$ Å, a configuration with an O–C–O bond angle of $107.5°$, gives us a converged result with C–O bond lengths of 1.17 Å and a O–C–O bond angle of $179.82°$. This geometry is extremely close to the one we found starting from a linear molecule. We should not expect the two geometries to be *exactly* the same; they were determined using iterative optimization methods that were halted once we were "sufficiently close" to the exact solution as dictated by the stopping criterion we used. It is quite reasonable to draw the conclusion from these calculations that our DFT calculations have predicted that CO_2 is a linear molecule (i.e., it has an $O-C-O$ bond angle of $180°$). Experimentally, CO_2 is known to be a linear molecule with $C-O$ bond lengths of 1.20 Å, so the DFT prediction is in good accord with experiment.

The ability to efficiently minimize the total energy of a collection of atoms is central to perhaps the majority of all DFT calculations. Before moving ahead, you should reread this section with the aim of summarizing the pitfalls we have identified in the two simple examples we examined. Developing a strong physical intuition about why these pitfalls exist and how they can be detected or avoided will save you enormous amounts of effort in your future calculations.

3.5.2 Geometry Optimization with Constrained Atoms

There are many types of calculations where it is useful to minimize the energy of a supercell by optimizing the position of some atoms while holding other atoms at fixed positions. Some specific examples of this situation are given in the next chapter when we look at calculations involving solid surfaces. Optimization problems involving constraints are in general much more numerically challenging than unconstrained optimization problems. Fortunately, performing a force-based geometry optimization such as the method outlined above is easily extended to situations where one or more of the atoms in a supercell is constrained. For the constrained system, only the positions of the unconstrained atoms are updated during each iteration of the optimization calculation. If the stopping criterion in a calculation of this kind is based on the magnitudes of forces on each atom, only the unconstrained atoms are included in defining the stopping criterion.

3.5.3 Optimizing Supercell Volume and Shape

The calculations above allowed the positions of atoms to change within a supercell while holding the size and shape of the supercell constant. But in the calculations we introduced in Chapter 2, we varied the size of the supercell to determine the lattice constant of several bulk solids. Hopefully you can see that the numerical optimization methods that allow us to optimize atomic positions can also be extended to optimize the size of a supercell. We will not delve into the details of these calculations—you should read the documentation of the DFT package you are using to find out how to use your package to do these types of calculations accurately. Instead, we will give an example. In Chapter 2 we attempted to find the lattice constant of Cu in the hcp crystal structure by doing individual calculations for many different values of the lattice parameters a and c (you should look back at Fig. 2.4). A much easier way to tackle this task is to create an initial supercell of hcp Cu with plausible values of a and c and to optimize the supercell volume and shape to minimize the total energy.* Performing this calculation using the same number of k points and so on as we used for hcp Cu in Chapter 2 gives an optimized structure with $a = 2.58$ Å and $c/a = 1.629$. These values are entirely consistent with our conclusions in Chapter 2, but they are considerably more precise, and we found them with much less work than was required to generate the data shown in Chapter 2.

EXERCISES

1. In the exercises for Chapter 2 we suggested calculations for several materials, including Pt in the cubic and fcc crystal structures and ScAl in the CsCl structure. Repeat these calculations, this time developing numerical evidence that your results are well converged in terms of sampling k space and energy cutoff.

2. We showed how to find a minimum of $f(x) = e^{-x}\cos x$ using the bisection method and Newton's method. Apply both of these methods to find the same minimum as was discussed above but using different initial estimates for the solution. How does this change the convergence properties illustrated in Fig. 3.6? This function has multiple minima. Use Newton's method to find at least two more of them.

3. Newton's method is only guaranteed to converge for initial estimates that are sufficiently close to a solution. To see this for yourself, try applying Newton's method to find values of x for which $g(x) = \tan^{-1} x = 0$. In this case, Newton's method is $x_{i+1} = x_i - (1 + x_i^2) \tan^{-1} x_i$. Explore how well this method converges for initial estimates including $x_0 = 0.1$, $x_0 = 1$, and $x_0 = 10$.

4. We did not define a stopping criterion for the multidimensional version of Newton's method. How would you define such a criterion?

5. Use methods similar to those in Section 3.5.1 to optimize the geometry of H_2O and hydrogen cyanide, HCN. HCN is a highly toxic gas that is nonetheless manufactured in large quantities because of its use as a chemical precursor in a wide range of industrial processes. Ensure that you have

---

*A subtlety in calculations where the volume of the supercell is allowed to change is that forces due to changes in supercell shape and volume can include systematic numerical errors unless the number of k points and energy cutoff are large. The artificial stress due to these effects is known as the Pulay stress (see Further Reading). One common approach to reducing this effect is to increase the energy cutoff by 30–50% during optimization of the supercell volume. If this is done, the total energy of the final optimized structure must be recalculated using the standard energy cutoff to complete the calculation.

sampled all relevant degrees of freedom by showing that multiple initial geometries for each molecule converge to the same final state. Compare your calculated geometries with experimental data.

6. In the exercises for Chapter 2, we suggested you compute the lattice constants, a and c, for hexagonal Hf. Repeat this calculation using an approach that optimizes the supercell volume and shape within your calculation. Is your result consistent with the result obtained more laboriously in Chapter 2? How large is the distortion of c/a away from the ideal spherical packing value?

7. Perform the calculations necessary to estimate the energy difference associated with forming an ordered bcc CuPd alloy from fcc Pd and fcc Cu. The ordered alloy is formed by defining a bcc crystal with Cu atoms at the corners of each cube and Pd atoms in the center of each cube (or vice versa). This ordered alloy is known to be the favored low temperature crystal structure of Pd and Cu when they are mixed with this stoichiometry. What does this observation tell you about the sign of the energy difference you are attempting to calculate? To calculate this energy difference you will need to optimize the lattice constant for each material and pay careful attention to how your energy cutoffs and k points are chosen.

REFERENCES

1. G. Kresse and D. Joubert, From Ultrasoft Pseudopotentials to the Projector Augmented-Wave Method, *Phys. Rev. B* **59** (1999), 1758.
2. S. R. Covey, *The Seven Habits of Highly Effective People*, Simon & Schuster, New York, 1989.

FURTHER READING

The concepts of reciprocal space, the Brillouin zone, and the like are staples in essentially all solid-state physics textbooks. The Further Reading sections in Chapters 1 and 2 list examples. Another source that gives a very clear introduction to the concepts of energy levels, energy bands, k space, and band structure is R. Hoffmann, C. Janiak, and C. Kollmar, *Macromolecules*, **24** (1991), 3725.

An excellent resource for learning about efficient numerical methods for optimization (and many other problems) is W. H. Press, S. A. Teukolsky, W. T. Vetterling, and B. P. Flannery, *Numerical Recipes in C++: The Art of Scientific Computing*, Cambridge University Press, Cambridge, UK, 2002. Multiple editions of this book are available with equivalent information in other computing languages.

A useful source for an in-depth discussion of the conjugate-gradient method is J. R. Shewchuk, An Introduction to the Conjugate Gradient Method Without the Agonizing Pain (http://www.cs.cmu.edu/~quake-papers/painless-conjugate-gradient.pdf).

The smearing method of Methfessel and Paxton is described in M. Methfessel and A. T. Paxton, *Phys. Rev. B* **40** (1989), 3616.

For details of tetrahedron methods, see P. E. Blöchl, O. Jepsen, and O. K. Andersen, *Phys. Rev. B* **49** (1994), 16223.

For more information on the history and details of pseudopotentials and the PAW method, see G. Kresse and D. Joubert, *Phys. Rev. B* **59** (1999), 1758, and the references therein.

For more information on Pulay stress and related complications associated with finite sets of plane waves and k points when calculating forces in supercells with varying volumes, see G. P. Francis and M. C. Payne, *J. Phys. Condens. Matter* **2** (1990), 4395.

APPENDIX CALCULATION DETAILS

All calculations in this chapter were performed using the PW91 GGA functional.

Sections 3.1.3 and 3.1.2 Bulk Cu calculations used a cubic supercell with 4 Cu atoms in the fcc structure and a cutoff energy of 292 eV. Methfessel–Paxton smearing with a width of 0.1 eV was used.

Section 3.5.1 Calculations for molecular N_2 and CO_2 used a cubic supercell of side length 10 Å, with reciprocal space sampled using $3 \times 3 \times 3$ k points placed with the Monkhorst–Pack method. The energy cutoff for these calculations was 358 eV.

4

DFT CALCULATIONS FOR SURFACES OF SOLIDS

4.1 IMPORTANCE OF SURFACES

Surfaces are technologically important in many fields, including catalysis, interfaces, membranes for gas separations, and semiconductor fabrication. Understanding the geometry and electronic structure of surfaces is important; for example, it has been established that there is often a correlation between the structure of a surface and its catalytic activity. One area of research in catalysis where DFT has played an important role is the effort to improve the technology and reduce the cost of the three-way catalysts that reduce CO, NO_x, and SO_x emissions from cars. These catalysts achieve the rather ambitious goal of oxidizing hydrocarbons and CO while simultaneously reducing NO_x. Traditionally, they have incorporated expensive materials such as platinum, and they have been subject to poisoning under certain conditions. DFT methods have played an important role in elucidating the mechanisms of the relevant reactions on catalytic metals, zeolites, and oxides and have led to improvements in the design of catalytic converters that improve their efficiency and lower their cost.

Surface science experiments and DFT have often been teammates in very successful projects. DFT has been used along with ultra-high-vacuum surface science experiments such as scanning tunneling microscopy (STM), temperature-programmed desorption, X-ray diffraction, and X-ray photoelectron spectroscopy

Density Functional Theory: A Practical Introduction. By David S. Sholl and Janice A. Steckel
Copyright © 2009 John Wiley & Sons, Inc.

to determine the surface structure of metals, metal oxides, nanoparticles, carbides, and sulfides. For example, in 1959 it was established by electron diffraction experiments that the Si(111) surface has a complex symmetry and that the arrangement of the atoms forming this surface must be very different from the bulk crystal structure of Si. Because of the technological importance of Si in microelectronics, there was great interest in understanding the details of this surface structure. It was not until 1992 that DFT calculations on a very large supercell allowed tests of the validity of what has since become accepted as the Si(111)-(7 × 7) surface reconstruction. These calculations not only defined the position of each atom in the surface, but they could be used to simulate how the surface would be imaged using STM, aiding in interpreting the beautiful but complex images seen experimentally.

Another technologically relevant material where DFT results have been coupled with a variety of experimental methods to produce a detailed understanding of chemistry on surfaces is titanium dioxide, TiO_2. Titanium dioxide is an important material used in pigments, oxygen sensors, antimicrobials and as a support for metal catalysts. Titanium dioxide also acts as a photosensitizer for photovoltaic cells and may be used as an electrode coating in photoelectrolysis cells, enhancing the efficiency of electrolytic splitting of water. DFT studies have been coupled with a wide variety of experimental techniques in order to characterize the binding of various atomic and molecular species with the surface as well as characteristics of the surface itself. The sites on the surface where oxygen atoms are missing are very important to a number of processes, and DFT calculations have not only helped to show that this is the case but to explain the physics that underlie this phenomenon.

In this chapter, we look at how DFT calculations can be used to examine surfaces of solids. After introducing the ideas necessary to define the atomic structure of bare surfaces, we give several examples of calculations relevant to gas–surface interfaces.

4.2 PERIODIC BOUNDARY CONDITIONS AND SLAB MODELS

If our goal is to study a surface, our ideal model would be a slice of material that is infinite in two dimensions, but finite along the surface normal. In order to accomplish this, it may seem natural to take advantage of periodic boundary conditions in two dimensions, but not the third. There are codes in which this technique is implemented, but it is more common to study a surface using a code that applies periodic boundary conditions in all three dimensions, and it is this approach we will discuss. The basic idea is illustrated in Fig. 4.1, where the supercell contains atoms along only a fraction of the vertical

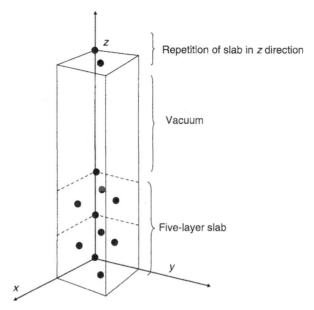

Figure 4.1 Supercell that defines a material with a solid surface when used with periodic boundary conditions in all three directions.

direction. The atoms in the lower portion of the supercell fill the entire super-cell in the x and y directions, but empty space has been left above the atoms in the top portion of the supercell. This model is called a slab model since, when the supercell is repeated in all three dimensions, it defines a series of stacked slabs of solid material separated by empty spaces, as shown schematically in Fig. 4.2. The empty space separating periodic images of the slab along the z direction is called the vacuum space. It is important when using such a model that there is enough vacuum space so that the electron density of the material tails off to zero in the vacuum and the top of one slab has essentially no effect on the bottom of the next. Figure 4.3 shows one view of the atoms in a material defined in this way seen from within the vacuum space. You will notice from this view that the supercell really defines two surfaces, an upper and lower surface.

Suppose we would like to carry out calculations on a surface of an fcc metal such as copper. How might we construct a slab model such as that depicted in Fig. 4.1? It is convenient to design a supercell using vectors coincident with the Cartesian x, y, and z axes with the z axis of the supercell coincident with the surface normal. Recall that for fcc metals, the lattice constant is equal to the length of the side of the cube of the conventional cell. The supercell vectors might then be

$$\mathbf{a}_1 = a(1,0,0), \qquad \mathbf{a}_2 = a(0,1,0), \qquad \mathbf{a}_3 = a(0,0,5), \qquad (4.1)$$

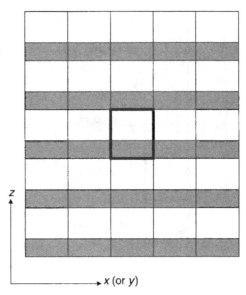

Figure 4.2   Schematic two-dimensional illustration of the material defined by the supercell in Fig. 4.1, showing 25 replicas of the supercell outlined by the thick lines. The shaded (white) regions indicate places in space occupied by atoms (vacuum).

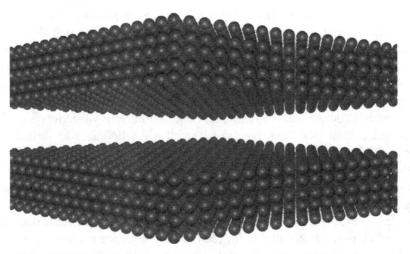

Figure 4.3   View of a five-layer slab model of a surface as used in a fully periodic calculation. In making this image, the supercell is similar to the one in Fig. 4.1 and is repeated 20 times in the x and y directions and 2 times in the z direction.

where a is the lattice constant. The fractional coordinates of the atoms in Fig. 4.1 are:

0.0, 0.0, 0.0	0.5, 0.5, 0.2
0.5, 0.5, 0.0	0.0, 0.5, 0.3
0.0, 0.5, 0.1	0.5, 0.0, 0.3
0.5, 0.0, 0.1	0.0, 0.0, 0.4
0.0, 0.0, 0.2	0.5, 0.5, 0.4

These choices define a slab with five distinct layers separated in the z direction by distance $a/2$ and a vacuum spacing of $3a$. Each layer in this slab has two distinct atoms within the supercell. The value of 5 in the definition of $\mathbf{a}_3$ is somewhat arbitrary. This vector must be large enough that there is a sufficient vacuum space in the supercell, but a larger vacuum spacing also means more computational effort is needed. In practice, you can determine how much vacuum is "enough" by checking the charge density in the vacuum space for several models with differing amounts of vacuum spacing. Ideally, the charge density will be close to zero in the vacuum space. If you try this yourself, be sure to note that if you change the length of $\mathbf{a}_3$ then the z components of the atoms' fractional coordinates must also be adjusted to maintain the correct physical spacing between the layers of the material. (If this sounds puzzling, this would be an excellent time to go back and review the definition of fractional coordinates.)

We would like our slab model to mimic the important characteristics of a real surface. The slab model discussed above includes five layers of an fcc metal. Of course, a real surface, except in some extreme situations, is the edge of a piece of material that is at least microns thick; this is nothing like our five-layer model at all. This begs the following question: how many layers are enough? Typically, more layers are better, but using more layers also inevitably means using more computational time. The number of layers that are deemed sufficient for convergence depends on the nature of the material studied and the property of interest. As was the case with the vacuum spacing, this question can be answered by carrying out calculations to see how some property (such as the surface energy or energy of adsorption) varies as the number of layers increases. We will see some examples of this idea below. Our choice for the number of layers in our model will usually be a compromise between computational cost and physical accuracy.

4.3 CHOOSING k POINTS FOR SURFACE CALCULATIONS

In Chapter 3, we discussed choosing k points for integrating over the Brillouin zone. The Monkhorst–Pack method for choosing k points may be used when choosing k points for slab calculations as well as for bulk calculations and many of the same considerations apply. To develop a little intuition about this task, it is useful to examine the reciprocal lattice vectors corresponding to the supercell defined above. Using Eq. (3.2), you can see that $|\mathbf{b}_1| = |\mathbf{b}_2| = 2\pi/a$ and $|\mathbf{b}_3| = 2\pi/5a$. Because our supercell has one "long" dimension in real space, it has one "short" dimension in reciprocal space. This suggests that we will not need as many k points in the $\mathbf{b}_3$ direction as in the other two directions to approximate integrals in reciprocal space. For the supercell we are currently discussing, using an $M_1 \times M_1 \times M_2$ k-point mesh, where $M_1 > M_2$ would seem appropriate from this geometric argument. This observation is similar to our discussion of choosing k points for hcp metals in Chapter 3.

The fact that the long dimension in the supercell includes a vacuum region introduces another useful observation for choosing k points. If the vacuum region is large enough, the electron density tails off to zero a short distance from the edge of the slab. It turns out that this means that accurate results are possible using just one k point in the $\mathbf{b}_3$ direction.[*] As a result, it is usual in slab calculations to use an $M \times N \times 1$ k-point mesh, where M and N are chosen to adequately sample k space in the plane of the surface. For the example we have defined above where $|\mathbf{b}_1| = |\mathbf{b}_2|$, M and N should be equal to match the symmetry of the supercell.

4.4   CLASSIFICATION OF SURFACES BY MILLER INDICES

If you look back at Fig. 4.3, you can imagine that one way to make the surfaces shown in that figure would be to physically cleave a bulk crystal. If you think about actually doing this with a real crystal, it may occur to you that there are many different planes along which you could cleave the crystal and that these different directions presumably reveal different kinds of surfaces in terms of the arrangement of atoms on the surface. This means that we need a notation to define exactly which surface we are considering.

As an example, Fig. 4.4 shows another view of the Cu surface that highlights where the plane of the surface is relative to the bulk crystal structure. The orientation of this plane can be defined by stating the direction of a vector normal to the plane.[†] From Fig. 4.4, you can see that one valid choice for this vector would be [0,0,1]. Another valid choice would be [0,0,2], or [0,0,7], and so on, since all of these vectors are parallel.

---

[*]This occurs because there is very little dispersion in the electronic band structure in this coordinate.
[†]The plane is usually identified by three indices enclosed in parentheses (hkl); the vector that is normal to the plane (in cubic systems) is enclosed in square brackets: $[hkl]$.

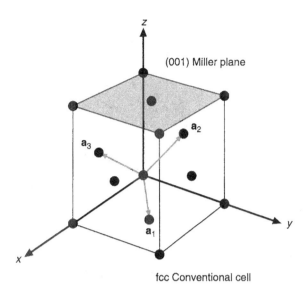

Figure 4.4 Conventional cell of an fcc metal with the (001) Miller plane highlighted.

To decide which of these many vectors to use, it is usual to specify the points at which the plane intersects the three axes of the material's primitive cell or the conventional cell (either may be used). The reciprocals of these intercepts are then multiplied by a scaling factor that makes each reciprocal an integer and also makes each integer as small as possible. The resulting set of numbers is called the *Miller index* of the surface. For the example in Fig. 4.4, the plane intersects the z axis of the conventional cell at 1 (in units of the lattice constant) and does not intersect the x and y axes at all. The reciprocals of these intercepts are $(1/\infty, 1/\infty, 1/1)$, and thus the surface is denoted (001). No scaling is needed for this set of indices, so the surface shown in the figure is called the (001) surface.

If the vector defining the surface normal requires a negative sign, that component of the Miller index is denoted with an overbar. Using this notation, the surface defined by looking up at the bottom face of the cube in Fig. 4.4 is the (00$\bar{1}$) surface. You should confirm that the other four faces of the cube in Fig. 4.4 are the (100), (010), ($\bar{1}$00), and (0$\bar{1}$0) surfaces. For many simple materials it is not necessary to distinguish between these six surfaces, which all have the same structure because of symmetry of the bulk crystal.[‡]

Figure 4.5 shows another surface that can be defined in a face-centered cubic material; this time the highlighted plane intercepts the x, y, and z axes at 1, 1, and 1. The reciprocals of these intercepts are $1/1$, $1/1$, and $1/1$.

[‡]The set of all six surfaces is denoted with curly brackets: {100}.

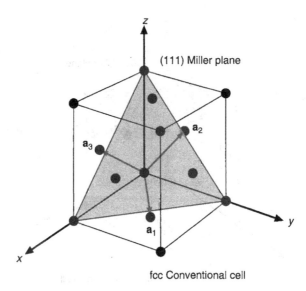

Figure 4.5   Similar to Fig. 4.4 but with the (111) Miller plane highlighted.

This surface is therefore the (111) surface. This surface is an important one because it has the highest possible density of atoms in the surface layer of any possible Miller index surface of an fcc material. Surfaces with the highest surface atom densities for a particular crystal structure are typically the most stable, and thus they play an important role in real crystals at equilibrium. This qualitative argument indicates that on a real polycrystal of Cu, the Cu(111) surface should represent a significant fraction of the crystal's surface total area.

The Cu(110) surface is depicted in Fig. 4.6. The plane of the (110) surface intersects the x and y axes at 1 and does not intersect the z axis at all. The reciprocals of these intercepts are $(1/1, 1/1, 1/\infty)$, and so the surface is designated the (110) Miller plane.

Figure 4.7 shows top-down views of the fcc (001), (111), and (110) surfaces. These views highlight the different symmetry of each surface. The (001) surface has fourfold symmetry, the (111) surface has threefold symmetry, and the (110) has twofold symmetry. These three fcc surfaces are all atomically flat in the sense that on each surface every atom on the surface has the same coordination and the same coordinate relative to the surface normal. Collectively, they are referred to as the low-index surfaces of fcc materials. Other crystal structures also have low-index surfaces, but they can have different Miller indices than for the fcc structure. For bcc materials, for example, the surface with the highest density of surface atoms is the (110) surface.

So far we have only shown examples of low-index surfaces. These surfaces are important in the real world because of their stability. They are also quite

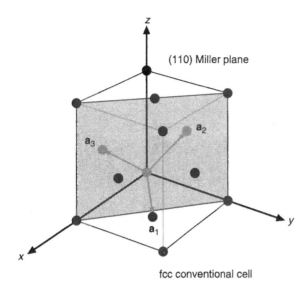

fcc conventional cell

Figure 4.6 Similar to Fig. 4.4 but with the (110) Miller plane highlighted.

convenient for DFT calculations because supercells defining low-index surfaces require only modest numbers of atoms in the supercell. There are many other surfaces, however, that have interesting and important properties. To give just one example, Fig. 4.8 shows top and side views of the Cu(322) surface. This surface has regions that locally look like the (111) surface, separated by steps one atom in height. The atoms located at step edges have a lower coordination number than other atoms in the surface, which often leads to high reactivity. Reactive step edge atoms play an important role in the catalytic synthesis of ammonia, which was described in the first vignette in Section 1.2. Although the structure of Cu(322) is more complicated than the low-index surfaces, it is still periodic in the plane of the surface, so it is possible to create a supercell that describes this surface.

Earlier, we mentioned that other crystal structures can have different Miller indices than the fcc structure. In the hexagonal close-packed (hcp)

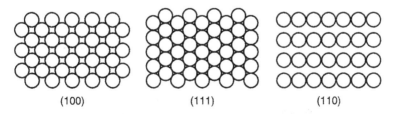

Figure 4.7 Top-down views of the (100), (111), and (110) surfaces of fcc metals. Only atoms in the top layer are shown.

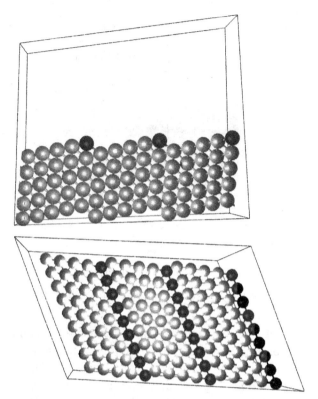

Figure 4.8   Top and side views of Cu(322) (upper and lower image, respectively). Three supercells are shown. In each image, the Cu atoms forming the step edge are dark gray.

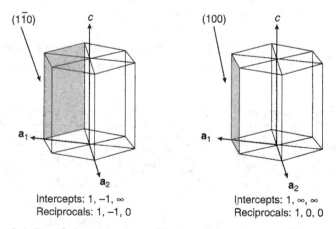

Figure 4.9   Labeling of planes in an hcp solid using a three-axis system. These two planes are equivalent by symmetry, but the indices obtained using three axes are not.

structure, spheres are arranged in a single close-packed layer to form a basal plane with each atom surrounded by six others. The next layer is added by placing spheres in alternating threefold hollows of the basal plane. If a third layer is added such that the spheres are directly above the spheres in the basal plane, we obtain the hcp structure. (If the third layer atoms are added in the hollows *not* directly above the basal plane, the fcc structure is obtained.)

Hexagonal close-packed (hcp) materials have a sixfold symmetry axis normal to the basal plane. Using a three-axis system to define Miller indices for this structure is unsatisfactory, as is demonstrated in Fig. 4.9. The two planes highlighted in Fig. 4.9 are equivalent by symmetry, and yet their Miller indices do not show this relationship. This is unfortunate since one of the reasons Miller indices are useful is that equivalent planes have similar

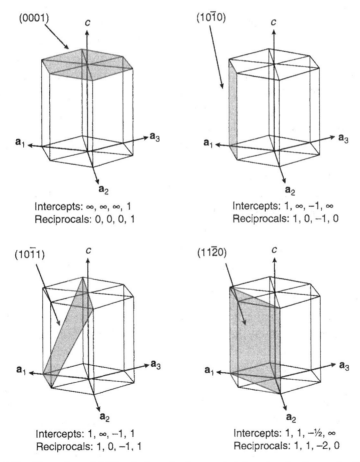

Figure 4.10   Examples of Miller indices for hcp materials using the four-axis, four-index system.

indices. Another shortcoming is that in this example the [*hkl*] direction is not normal to the (*hkl*) plane.

A better solution is to use a four-axis, four-index system for hcp solids. A few examples are depicted in Fig. 4.10. The Miller indices are found as before by taking the reciprocals of the intercepts of the plane with the four axes. Using this system, the equivalent planes discussed above become ($1\bar{1}00$) and ($10\bar{1}0$). Now the six equivalent planes resulting from the sixfold axis of symmetry can be identified as the {$1\bar{1}00$} family of planes. Using four axes, the [*hkil*] direction is normal to the (*hkil*) plane, in the same way it was for cubic solids using the three-axis system.[§]

4.5   SURFACE RELAXATION

In the example above, we placed atoms in our slab model in order to create a five-layer slab. The positions of the atoms were the ideal, bulk positions for the fcc material. In a bulk fcc metal, the distance between any two adjacent layers must be identical. But there is no reason that layers of the material near a surface must retain the same spacings. On the contrary, since the coordination of atoms in the surface is reduced compared with those in the bulk, it is natural to expect that the spacings between layers near the surface might be somewhat different from those in the bulk. This phenomenon is called surface relaxation, and a reasonable goal of our initial calculations with a surface is to characterize this relaxation.

In Fig. 4.11, the positions of atoms in a DFT calculation of a five-layer slab model of a surface before and after relaxation are indicated schematically. On the left is the original slab model with the atoms placed at bulk positions and on the right is the slab model after relaxation of the top three layers. Surface relaxation implies that the relaxed surface has a lower energy than the original, ideal surface. We can find the geometry of the relaxed surface by performing an energy minimization as a function of the positions of the atoms in the super-cell. We imagine the bottom of the slab as representing the bulk part of a material and constrain the atoms in the bottom layers in their ideal, bulk positions. The calculation then involves a minimization of the total energy of the supercell as a function of the positions of the atoms, with only the atoms in the top layers allowed to move, as described in Section 3.5.2. This results in a structure such as the one shown on the right in Fig. 4.11. In reality, the movements of the atoms in surfaces are on the order of 0.1 Å. In this figure, the magnitude of the relaxation of the top layers is exaggerated.

[§]Because the four axes are not independent, the first three indices in the four-index system will always sum to zero. (This serves as a convenient check.)

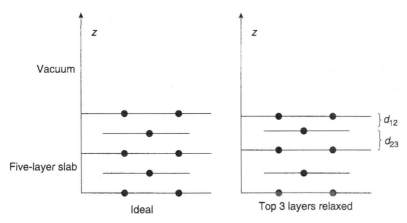

Figure 4.11 Schematic illustration of relaxation of surface atoms in a slab model. The top three layers of atoms were allowed to relax while the bottom two layers were held at the ideal, bulk positions.

The abrupt termination of a surface in the direction of the surface normal leads to a dramatic change in coordination of the atoms at the surface. Often this leads to a decrease in the distance between the first and second atomic layers. The distance between the outermost surface layer and the second layer can be denoted d_{12} and the change upon relaxation is denoted δd_{12}. The change upon relaxation is often expressed as a percent of the bulk material layer spacing, with negative numbers indicating a contraction and positive numbers indicating expansion between the layers. The changes in the distance between other layers are denoted δd_{23}, δd_{34}, and so forth. Table 4.1 lists interlayer spacing relaxations calculated for the Cu(100) and (111) surfaces using slab models with different numbers of layers. In each case, the two bottom layers were fixed at bulk positions while all other layers were allowed to relax.

TABLE 4.1 Interlayer Relaxations in Cu(100) and Cu(111) Calculated Using DFT as Function of Slab Thickness

	Cu(001)			Cu(111)		
	δd_{12} (%)	δd_{23} (%)	δd_{34} (%)	δd_{12} (%)	δd_{23} (%)	δd_{34} (%)
5 layers	−3.84	−0.50	−0.53	−0.61	−0.08	+0.08
6 layers	−1.93	+0.83	+0.37	−0.64	−0.11	+0.27
7 layers	−2.30	+0.55	−0.25	−0.56	−0.04	+0.32
8 layers	−2.14	+0.85	+0.00	−0.59	−0.32	+0.51
Expt.	−2.0 ± 0.5[a]	+1.0 ± 0.7[a]		−0.7 ± 0.5[b]		

[a]From Ref. 1, measured at 305 K.
[b]From Ref. 2.

The changes in interlayer spacing are more dramatic on the Cu(100) surface, while on the close-packed Cu(111) surface they are less pronounced. For both surfaces, the calculated change in interlayer spacing converges toward the experimental result as more layers are incorporated into the model. Using the eight-layer slab, the calculated relaxation for δd_{12} on Cu(100) is -2.14%, or about 0.036 Å. This is within the error bars on the experimental value of $-2.0 \pm 0.5\%$.

Likewise, for the Cu(111) surface, the DFT estimate for δd_{12} is -0.59%, or 0.012 Å; this value is very close to the experimental number. Considering how small these changes in position are, DFT is quite successful in reproducing the experimental measurements.

In our discussion of surface relaxation, we have only considered relaxation in the direction normal to the surface. This is the correct description of our calculation because the slab is completely symmetric in the plane of the surface. As a result, the components of force acting on each atom in the plane of the surface are precisely zero, and the atoms cannot move in this plane during relaxation. This is also the physically correct result for a real surface; it is not just an artifact of our calculation. If we view the dimensions of the supercell in the plane of the surface as being dictated by these bulklike layers, then it is clear that our supercell must be defined using the bulk lattice spacing, as was done in Eq. (4.1). This discussion highlights one more feature of surface calculations we have not yet mentioned, namely that we must choose what numerical value of the lattice constant to use. The best approach is to use the lattice constant obtained from a carefully converged DFT calculation for the bulk material using the same exchange–correlation functional that is to be used for the surface calculations. Using any other lattice constant (e.g., the experimental lattice constant) leads to artificial stresses within the material that result in physically spurious relaxation of the surface atoms.

4.6   CALCULATION OF SURFACE ENERGIES

As we have already discussed, surfaces can be created by cleavage along some plane in the bulk material. The surface energy, σ, is the energy needed to cleave the bulk crystal. We can calculate this energy by realizing that the energy associated with the cutting process is equal to the energy of the two surfaces that were created (if the process is reversible). This implies that the surface energy can be determined from a DFT slab calculation using

$$\sigma = \frac{1}{A}\left[E_{\text{slab}} - nE_{\text{bulk}}\right], \tag{4.2}$$

TABLE 4.2   Surface Energies Calculated for Cu(100) and Cu(111) from DFT as Function of Slab Thickness, in eV/Å^2 (J/m^2)^a

Slab Model	σ, Cu(100)	σ, Cu(111)
5 layers	0.094 (1.50)	0.087 (1.40)
6 layers	0.097 (1.55)	0.089 (1.43)
7 layers	0.098 (1.57)	0.089 (1.43)
8 layers	0.096 (1.53)	0.091 (1.46)
Expt.		0.114 (1.83)^b

^{a}In each calculation, the bottom two layers were constrained in their bulk positions and all other layers were allowed to relax.
^{b}From Ref. 3.

where E_{slab} is the total energy of the slab model for the surface, E_{bulk} is the energy of one atom or formula unit of the material in the bulk, n is the number of atoms or formula units in the slab model, and A is the total area of the surfaces (top and bottom) in the slab model.[||] Macroscopically, surface energy is typically expressed in units of joules/meters squared (J/m^2). In DFT calculations it is more natural to define surface energies in electron volts/angstroms squared (eV/Å^2), so it is convenient to note that $1 \text{ J/m}^2 = 16.02 \text{ eV/Å}^2$.

The surface energy defined in Eq. (4.2) is the difference of two quantities that are calculated in somewhat different fashions. In the case of the surface, one would typically be using a comparatively large supercell, including a vacuum space, and using comparatively few k points. In the case of the bulk, the opposite would be true. How, then, can we be sure that the difference in the theoretical treatments does not influence our answer? Unfortunately, there is no single solution to this problem, although the problem can be minimized by making every effort to ensure that each of the two energies are well converged with respect to number of layers in the slab model, k points, energy cutoff, supercell size, and the like.

As an example, the DFT-calculated surface energies of copper surfaces are shown in Table 4.2 for the same set of slab calculations that was described in Table 4.1. The surface energy of Cu(111) is lower than for Cu(100), meaning that Cu(111) is more stable (or more "bulklike") than Cu(100). This is consistent with the comment we made in Section 4.4 that the most stable surfaces of simple materials are typically those with the highest density of surface atoms. We can compare our calculated surface energy with an experimental

---

[||]This definition neglects entropic contributions to the surface energy.

observation for this quantity. It is clear from this comparison that the DFT calculations we have outlined give a good account of this surface energy.

A striking difference between the results in Tables 4.1 and 4.2 is that while the former suggested that one must include eight layers to achieve a well-converged result for the change upon relaxation δd_{12}, the latter suggests that reasonable estimates for the surface energy can be obtained using a four-layer slab. This is an illustration of the important idea that "convergence" is not a concept that applies equally to all possible physical properties. It is quite possible to have a set of calculations that is "well converged" (i.e., provides a numerically stable result that will not change if higher resolution calculations are performed) for one property but does not fall into this category for another property. In the calculation of the relaxation of the interlayer spacing, a change in spacing of only 0.02 Å results in a δd_{12} of 1%. This means that this quantity is a very sensitive probe of the precise geometry of the atoms in the surface. The surface energy results in Table 4.2, in contrast, show that this quantity is less sensitive to the precise geometry of the surface layers. This discussion gives one example of why it is best to think of calculations being well converged with respect to a particular property, rather than just well converged in general.

4.7 SYMMETRIC AND ASYMMETRIC SLAB MODELS

Let us return to our five-layer slab of fcc material that was introduced in Fig. 4.1. When we allowed this model to relax, we chose to relax the top three layers and keep the bottom two layers fixed at bulk positions. This approach defines an asymmetric slab; layers on one side are relaxed to mimic the surface and layers on the other side are kept fixed to mimic the bulk region. An important feature of an asymmetric slab model is that surface processes of interest may generate a dipole. This is particularly important if we wish to examine the adsorption of atoms or molecules on a surface. For example, imagine that one would like to examine the adsorption of fluorine on a surface. The dipole created by adsorption of such an electronegative atom could be sizable. When periodic boundary conditions are applied to a supercell containing this nonzero dipole, a contribution to the total energy from the electrostatic interaction of this dipole with other images of the dipole above and below it arises. This electrostatic interaction is a mathematically correct feature in a periodic system, but it is physically spurious if we are really trying to describe an isolated surface. Many DFT codes have implemented a scheme for canceling the artificial electrostatic field generated by the application of periodic boundary conditions to a model with a nonzero dipole. Typically, the dipole can be calculated and an equal and opposite dipole applied as a correction to the local potential in the vacuum region.

An alternative is to describe the surface using a symmetric model. In the symmetric model, the center of the slab consists of a mirror plane. The atoms in the middle layers are typically fixed at bulk geometries and the layers above and below are allowed to relax. One advantage of a symmetric model is that any dipole generated by surface features will be automatically canceled. There is a cost involved, however, because it is typically necessary to include more layers in a symmetric slab than in an asymmetric slab. A symmetric slab with nine layers is depicted in Fig. 4.12. Note that, in this example, three layers are allowed to relax on each side of the slab. Recall that in our earlier model of an asymmetric slab, three layers were allowed to relax on one side of a five-layer slab. So in order to carry out calculations in which three layers are allowed to relax, one would need to employ nine layers in a symmetric model, compared to the five layers needed for an asymmetric one.

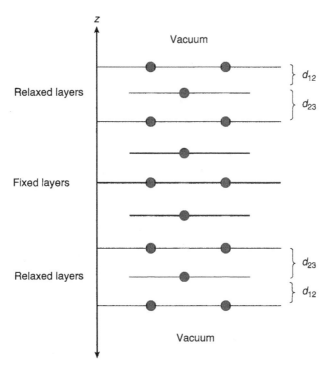

Figure 4.12   Schematic illustration of symmetric surface slab containing nine layers. With the middle three layers fixed and the outer layers relaxed, the interlayer spacings (and all other properties) on the bottom and top of the slab must be identical.

4.8   SURFACE RECONSTRUCTION

Earlier in this chapter, we mentioned that the atoms forming surfaces differ from atoms in a bulk material because of surface relaxation. It is important to recognize, however, that DFT calculations may not yield accurate information about a surface merely by allowing a surface to relax. Numerous surfaces undergo *reconstructions* in which surface atoms form new bonds.[#]

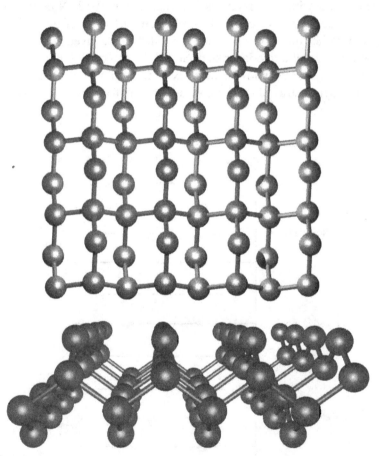

Figure 4.13   Top and side view of unreconstructed Si(001) as terminated from the bulk Si structure. Coordinates used for this illustration come from a fully relaxed DFT calculation.

[#]Another way to describe surface reconstruction is that during a reconstruction the symmetry of the surface atoms change in some way relative to their symmetry in the (relaxed) bulk-terminated surface.

A very clear example of a surface reconstruction is given by the Si(100) surface. Silicon, carbon, and some other Group IV materials adopt the diamond structure in their bulk form. In this structure, each atom is bonded in a tetrahedral arrangement with its four nearest neighbors. If we cleave silicon along the (001) Miller plane, we are left with silicon atoms at the surface that have two nearest neighbors instead of four, as illustrated in Fig. 4.13. This is a perfectly well-defined material and we can use DFT to relax the surface atoms just as we did for Cu surfaces before. The images in Fig. 4.13 were generated from calculations of this kind.

From a chemical perspective, the bulk termination of Si(001) is not entirely satisfactory because each surface silicon atom has two dangling bonds associated with an unpaired electron. You can imagine that a surface atom might prefer to move into a position where unpaired electrons could pair with other unpaired electrons and form new bonds, even though doing so will

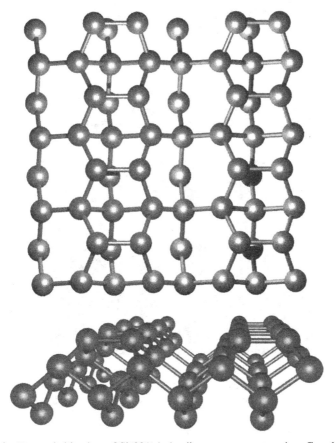

Figure 4.14   Top and side view of Si(001) in its dimer row reconstruction. Coordinates used for this illustration come from a fully relaxed DFT calculation.

cost some energy in terms of placing new strains on the existing bonds in the surface. In this example, the energy gained by connecting dangling bonds is so large that this is exactly what happens in nature. The atoms in the surface can get into a much more energetically stable state by pairing up. Figure 4.14 shows top and side views of the reconstructed Si(100) surface in which neighboring surface atoms have moved closer to one another and formed new bonds. The energy difference between the DFT structures used in Figs 4.13 and 4.14, both of

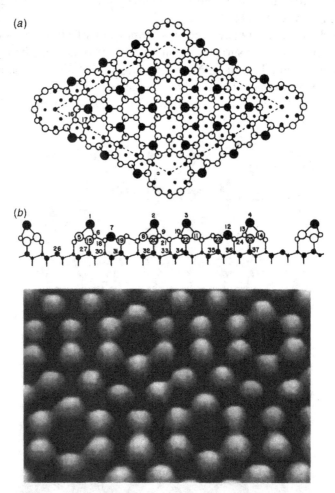

Figure 4.15   (a) Diagram of the Si(111)-7 × 7 reconstruction on Si(111) from the work of Brommer et al. Large solid circles denote adatoms. The size of the circle indicates the distance of the atoms from the top of the surface. The (7 × 7) unit cell is outlined. (b) STM image of the reconstruction. [Reprinted with permission from K. D. Brommer, M. Needels, B. E. Larson, and J. D. Joannopoulos, Ab Initio Theory of the Si(111)-(7×7) Surface Reconstruction: A Challenge for Massively Parallel Computation, *Phys. Rev. Lett.* **68** (1992), 1355 (Copyright 1992 by the American Physical Society).]

which were fully relaxed, is ~ 0.7 eV/surface atom. This is a very large amount of energy! The pairs of atoms on the surface form long rows called dimer rows. These distinctive rows are clearly visible in STM images of Si(001).

Once you appreciate that surface reconstructions exist in nature, the key point to take away from this discussion is that the Si(001) reconstruction shown in Fig. 4.14 does *not* emerge spontaneously if a DFT calculation is performed using the bulk termination of Si for the starting geometry. In the language of optimization, the reconstructed surface defines the *global* minimum in the energy of the surface, but the bulk termination of the surface leads to a different *local* minimum in energy. Because DFT calculations only locate local minima in energy, no set of DFT calculations can be used to "prove" that a surface reconstruction does not occur for some surface of interest. This situation is analogous to our discussion of "predicting" crystal structures in Section 2.4.

As a cautionary example, consider another surface of Si, the (111) surface that was mentioned in the introduction to this chapter. Using the same chemical reasoning as above, it is not hard to believe that this surface will reconstruct in some way. If we were aiming to "predict" the details of this reconstruction, we could perform a series of calculations that connect together dangling bonds on the surface in various possible combinations. After doing multiple calculations of this kind, whichever structure had the lowest energy per surface atom would be our "prediction." We could first do this in calculations with a single unit cell of the Si(111) surface, and if we were more ambitious we might try examples with more than one unit cell. In reality, the stable reconstruction of Si(111) involves the intricate arrangement of atoms formed from a 7×7 set of unit cells (see Fig. 4.15). DFT played an important role in showing that this structure is the correct one, but only after a great deal of experimental data had been collected to guide these calculations.

4.9 ADSORBATES ON SURFACES

So far in this chapter we have only looked at clean surfaces, that is, surfaces with the same chemical composition as the bulk material from which they are made. Many of the interesting things that happen with surfaces, of course, occur when other chemicals are present on them, so we now turn to that situation. We will only look at situations where atoms or molecules are chemisorbed on surfaces—this can loosely be defined as the situations where a chemical bond of some kind exists between the atom or molecule and the surface.

As an example, let us perform some calculations to understand how H atoms bind on Cu(100), the metal surface we looked at earlier in the chapter. Many chemical reactions on metals involve the creation of individual H atoms

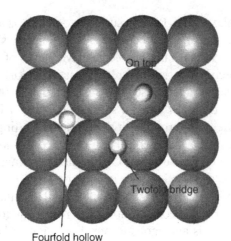

Figure 4.16 Illustration of potential high symmetry binding sites of H (filled circles) on Cu(100). Atoms in topmost layer of the Cu(100) surface are shown as large open circles.

(among other chemical species), so knowing the characteristics of these atoms on a surface would be a necessary step toward describing any of these reactions. Perhaps the simplest question we can ask is: where on the surface does H prefer to be? If we look at the (100) surface, there are several sites with special symmetry that are intuitively appealing as potential binding sites for H. These are illustrated in Fig. 4.16.

You can probably now guess how to perform a calculation to describe H in each of the positions defined in Fig. 4.16. We define a supercell containing a slab model of Cu(100), place an H atom relatively close to the top layer of the surface in one of the positions shown in the figure, and then minimize the system's energy while allowing several of the surface layers and the H atom positions to relax. A simple way to interpret a set of calculations like this is to simply compare their total energies, since the differences in their total energies correspond to the energy cost (or gain) in moving H between the various sites on the surface. Results from a set of calculations of this kind are shown in Table 4.3. These calculations predict that H is most stable in the fourfold

TABLE 4.3 Results of Calculations for H Adsorption on Cu(100) as Described in Text[a]

Initial H Site	Adsorption Energy Relative to Hollow Site	Adsorption Energy Relative to $H_2(g)$	Local Minimum?
Fourfold hollow	0	−0.19	Yes
Twofold bridge	0.08	−0.11	No
On top	0.57	+0.38	No

[a]All energies are in electron volts (eV).

hollow site, but that the twofold bridge site is only 0.08 eV higher in energy. Since a typical unit of thermal energy at room temperature is $k_b T \sim 0.03$ eV, these results suggest that H might be found in both of these sites under at least some circumstances.

Before we finish thinking about these results, it is a good idea to verify that the atomic configurations that have been found come from true local energy minima for the system. To see why this might matter, look for a moment at the on-top site in Fig. 4.16. The force in the plane of the surface on an H atom that lies precisely above a surface Cu atom must be precisely zero, *by symmetry*. This means that during our optimization calculation, the H atom will stay exactly above the Cu atom even if its energy could be lowered by moving away from this symmetric location. In fact, this argument can be applied to all three of the high-symmetry sites identified in the figure. We saw a similar situation when we considered the geometry of CO_2 in Section 3.5.1.

We can avoid this symmetry-induced trap by deliberately breaking the symmetry of our atom's coordinates. One easy way to do this is to repeat our calculations after moving the H atom a small amount (say, 0.2 Å) in some arbitrary direction that does not coincide with one of the symmetry directions on the surface. What we find, if we run calculations in which we start the H atom at a point about 0.2 Å away from each of the high-symmetry sites mentioned above is that the H atom relaxes to the fourfold hollow site even if it is started quite near the top and bridge sites. This shows that the top and bridge sites are not minima for this system.

Our calculations so far tell us where on the Cu(100) surfaces H atoms prefer to be. Another simple but important question is: how strongly do the atoms prefer to be on the surface instead of somewhere else completely? This question is usually answered by calculating the adsorption energy of the species on the surface. One possible definition for this quantity is

$$E_{ads}^{atomic} = E_{H/surf} - E_{H(g)} - E_{surf}. \tag{4.3}$$

Here, the three terms on the right are the total energy of surface with H adsorbed on it, the total energy of a single H atom by itself in the gas phase, and the total energy of the bare surface. Simply put, this quantity is the amount of energy required to pull an H atom off the surface into the gas phase. This definition is easy to use, but chemically unnatural because H atoms rarely if ever exist by themselves for very long. A more physically meaningful quantity is the energy gained (or lost) by pulling two H atoms off the surface and forming an H_2 molecule in the gas phase. This is

$$E_{ads} = E_{H/surf} - \tfrac{1}{2} E_{H_2(g)} - E_{surf}. \tag{4.4}$$

This adsorption energy is very different from the value defined in Eq. (4.3) because of the significant bond energy in H_2 molecules. The energy of adsorption for H on the Cu(100) surface relative to half of the H_2 molecule as defined by Eq. (4.4) is given in the right-hand column in Table 4.3. A negative number signifies that adsorption is lower in total energy compared to the bare surface and half the gas-phase H_2 molecule. It can be seen that the process is thermodynamically favored for the fourfold site. Although the other two sites we examined on the surface are not local minima for adsorbed H atoms, we can still describe their energy using Eq. (4.4). The result for the bridge site is negative, but it is only just barely so. The result for the top site, however, is large and positive, meaning that even if this site was a local minimum, the number of H atoms that would be observed in this site if the surface was in equilibrium with gaseous H_2 would be extremely small.

The ideas above extend in a natural way to probing the adsorption of molecules on surfaces. If you wanted to describe adsorption of hydroxyls (OH) on Cu(100), you would need to perform separate calculations for a range of possible binding sites for these species on the surface. Unlike our calculations for atomic H, these calculations would need to explore both the position of the species bonding to the surface and also the orientation of the O–H bond relative to the surface. Similar to atomic adsorption, a common misstep for simple species such as OH is to use calculations that only explore highly symmetric configurations such as those with the O–H bond perpendicular to the surface. As the size of the molecule adsorbing on a surface increases, the number of possible ways that molecular adsorption can occur grows rapidly. In many situations where the molecule of interest contains more than a handful of atoms, it is difficult to conclusively determine the favored binding configuration without experimental guidance.

4.9.1   Accuracy of Adsorption Energies

We have now characterized the binding of H on Cu(100) in two ways, with the adsorption energy relative to formation of gaseous H_2 and (in Table 4.3) with relative adsorption energies among surface states. The accuracy that can be expected with DFT for these two types of quantities is quite different. As discussed in Chapter 1, DFT calculations make systematic (but unknown) errors relative to the true solution of the Schrödinger equation because of the approximate nature of the exchange–correlation functional. A very broad body of calculations has shown, however, that: *if two DFT calculations compare "chemically similar" states, then the systematic errors in these two calculations are also similar.* In other words, relative energies of chemically similar states can be calculated much more accurately with DFT than absolute energies.

Adsorption on surfaces gives us a concrete example of this general principal. If we calculate the relative energy of H in a fourfold and a twofold site, then it is certainly reasonable to describe these as chemically similar. We can therefore expect a good level of accuracy in this relative energy. An H atom adsorbed on a metal surface and an H atom in a gaseous H_2 molecule, by contrast, cannot be described as being chemically similar. We can, therefore, expect that the error in the adsorption energy defined by Eq. (4.4) as computed with DFT will be larger than the error in the relative energies of the different surface states. We return to this point as part of a more detailed discussion of the precision of DFT calculations in Chapter 10.

4.10 EFFECTS OF SURFACE COVERAGE

In Table 4.3, we listed the calculated adsorption energy of H on the Cu(100) surface (relative to gaseous H_2). However, when we presented this number, we glossed over a potentially important contribution to the adsorption energy, namely the effects of surface coverage. Because we are using periodic boundary conditions, putting one adsorbate in our supercell automatically means that each adsorbate "sees" a copy of itself in each of the neighboring supercells. The crux of the matter boils down to the following point: in setting up a supercell with an adsorbate on a surface, we are necessarily describing a periodic, regularly repeated surface overlayer.

There are two issues that are relevant here. First, the adsorbates in a supercell calculation necessarily have a long-range pattern that is repeated exactly as the supercell is repeated. With periodic boundary conditions, it is impossible to model any kind of truly random arrangement of adsorbates. The good news is that in nature, it happens that adsorbates on crystal surfaces often do exhibit long-range ordering, especially at low temperatures, so it is possible for calculations to imitate real systems in many cases.

The second issue is that the size of the supercell controls the distance between adsorbates. If the supercell is small, it defines a surface with a high density (or coverage) of adsorbates. If the supercell is large, surfaces with lower coverages can be defined. When there is one adsorbate per surface atom, the adsorbed layer is often said to have a coverage of 1 monolayer (ML). If there is one adsorbate per two surface atoms, the coverage is 0.5 ML, and so on.**

As you might imagine, a system of naming has evolved for discussing the symmetry of overlayers. Instead of delving into the rationale behind all the names, we simply give some representative examples.

**The nomenclature is not so clearly defined for surfaces that have more than one kind of surface atom, such as surfaces of alloys or compounds.

In Fig. 4.17, several examples of H adsorption on Cu(100) are depicted. It may be helpful to start with three examples. Going from Fig. 4.17(a) to (b) to (d), the coverage decreases from 1 ML to a 0.5–0.125 ML. The names for these overlayers are (1×1), $c(2 \times 2)$, and $c(4 \times 4)$. If you look at the length of the sides of the supercells as they are depicted in the figures and compare that to the number of adsorbates along each side, it might seem logical that these cells are named in this fashion. The "c" in these names stands for "centered," meaning that the supercell has an adsorbate atom in its center as well as at its corners. In our example, the H atoms are positioned at the hollow sites, but they could all have been placed at top sites or bridge sites, or any other site, and the overlayers would have been named in exactly the same way. If you now compare Fig. 4.17b and c, you see that these both depict the $c(2 \times 2)$ overlayer. The difference between these two representations is that the supercell in Fig. 4.17c is quite a bit smaller and thus will require a smaller computational investment. The supercell in Fig. 4.17b was presented to help

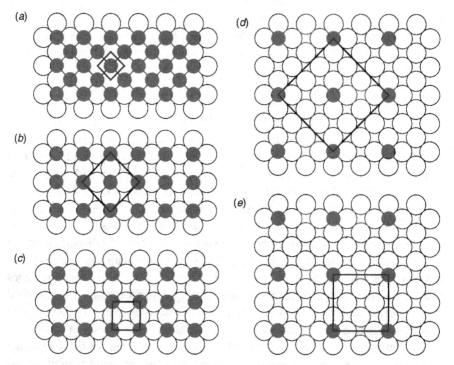

Figure 4.17 Examples of H adsorption on Cu(100) at different coverages. (a) c(1 × 1) ordering with coverage 1 ML; (b) c(2 × 2) ordering with coverage 0.5 ML; (c) the same overlayer as (b), but constructed using a smaller supercell; (d) c(4 × 4) ordering with 0.125 ML; (e) the same overlayer as (d), but with a smaller supercell. Cu atoms are open circles, adsorbed H atoms are depicted as gray circles, and a supercell is indicated with black lines.

TABLE 4.4   Energy of Adsorption of H on Cu(100) for Three Different Surface Overlayers

	(1×1)	$c(2 \times 2)$	$c(4 \times 4)$
H coverage	1.00 ML	0.50 ML	0.125 ML
$E_{(ads)}$ relative to $\frac{1}{2}H_2$ (eV)	-0.08	-0.11	-0.19

show why the $c(2 \times 2)$ name is used for this overlayer. A similar relationship exists between the overlayers in Fig. 4.17d and e.

Now let us return to the question of whether the coverage makes a difference in our computed adsorption energy. The adsorption energies computed with DFT for H atoms in the hollow sites on Cu(100) at several different coverages are listed in Table 4.4. If you suspected that leaving more space between adsorbed H atoms would lower the energy of adsorption, you were right. The energy of adsorption for the $c(4 \times 4)$ overlayer, which is the lowest coverage modeled here, is the lowest; it is the most favorable. The (1×1) overlayer in which an H atom occupies every hollow site on the (100) surface is barely thermodynamically favored relative to gaseous H_2. For most calculations, we cannot ignore the effects of neighboring adsorbates. This simple example points out that in describing the adsorption energy of an atom or a molecule

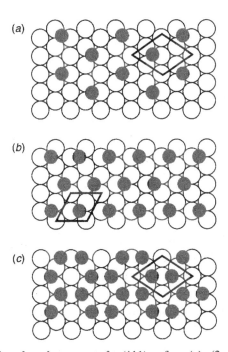

Figure 4.18   Examples of overlayers on an fcc (111) surface. (a) p(2 × 2) 0.25 ML overlayer (b) ($\sqrt{3} \times \sqrt{3}$)R30° 0.333 ML; (c) p(2 × 2) 0.50 ML overlayer.

on a surface, it is important to define the coverage and symmetry of the adsorbates in order to completely define the system being considered.

Some examples of overlayers on an fcc (111) surface are presented in Fig. 4.18. The 'p' in the names of these overlayers denotes the unit cells of the overlayers as 'primitive'. Note that for the close-packed (111) surface, full monolayer coverage is even less energetically favorable than it is on the more open (100) surface because adsorbates would be forced into such close quarters.

EXERCISES

In all of these exercises, be careful to consider aspects of numerical convergence related to the number of surface layers, the sampling of k space, and the like. As you gain experience from the first few exercises, fewer calculations to test convergence will be needed in the later calculations if you can clearly explain how the convergence information developed in the earlier exercises is relevant for the situations in the later exercises.

1. Develop supercells suitable for performing calculations with the (100), (110), and (111) surfaces of an fcc metal. What size surface unit cell is needed for each surface to examine surface relaxation of each surface?
2. Extend your calculations from Exercise 1 to calculate the surface energy of Pt(100), Pt(110), and Pt(111).
3. Pt(110) is known experimentally to reconstruct into the so-called missing-row reconstruction. In this reconstruction, alternate rows from the top layer of the surface in a (2 × 1) surface unit cell are missing. Use a supercell defined by a (2 × 1) surface unit cell of Pt(110) to compute the surface energy of the unreconstructed and reconstructed surfaces. Why does this comparison need to be made based on surface energy rather than simply based on the total energy of the supercells? Are your results consistent with the experimental observations? Use similar calculations to predict whether a similar reconstruction would be expected to exist for Cu(110).
4. Perform calculations to determine the preferred binding site for atomic O on Pt(111) using ordered overlayers with coverages of 0.25 and 0.33 ML. Does the adsorption energy increase or decrease as the surface coverage is increased?
5. Perform calculations similar to Exercise 4, but for the adsorption of hydroxyl groups, OH, on Pt(111). What tilt angle does the OH bond form with the surface normal in its preferred adsorption configuration? What numerical evidence can you provide that your calculations adequately explored the possible tilt angles?

6. OH groups could appear on Pt surfaces due to the dissociation of water molecules on the surface. To estimate whether this process is energetically favorable, compare the energy of clean Pt(111) surface and a gas-phase H_2O molecule to the energy of an OH group coadsorbed on Pt(111) with an H atom.

REFERENCES

1. D. E. Fowler and J. V. Barth, Structure and Dynamics of the Cu(001) surface Investigated by Medium-Energy Ion Scattering, *Phys. Rev. B* **52** (1995), 2117.
2. S. Å. Lindgren, L. Walldén, J. Rundgren, and P. Westrin, Low-Energy Electron Diffraction from Cu(111): Subthreshold Effect and Energy-Dependent Inner Potential; Surface Relaxation and Metric Distances between Spectra, *Phys. Rev. B* **29** (1984), 576.
3. F. R. de Boer, R. Boom, W. C. M. Mattens, A. R. Miedema, and A. K. Niessen, *Cohesion in Metals*, North-Holland, Amsterdam, 1988.

FURTHER READING

Two books that give a detailed account of the notation and physical phenomena associated with crystalline surfaces are J. B. Hudson, *Surface Science: An Introduction*, Butterworth-Heinemann, Boston, 1992, and G. A. Somorjai, *Introduction to Surface Chemistry and Catalysis*, Wiley, New York, 1994.

There is an enormous literature describing DFT calculations relevant to surface processes. Here we list several historical and representative examples.

For details on the reconstruction of Si(111), see K. D. Brommer, M. Needels, B. E. Larson, and J. D. Joannopoulos, *Phys. Rev. Lett.* **68** (1992), 1355.

For an example of combining insights from DFT calculations and experiments to understand oxygen interactions with the technologically important TiO_2 surface see: M. A. Henderson, W. S. Epling, C. L. Perkins, C. H. F. Peden, and U. Diebold, *J. Phys. Chem. B* **103** (1999), 5328, and references therein. A related example of using DFT to examine materials relevant to improving pollution control in automobiles is W. F. Schneider, J. Li, and K. C. Hass, *J. Phys. Chem. B* **105** (2001), 6972.

For an interesting discussion of the calculation of surface energies, as well as a discussion of the accuracy of LDA, GGA, and several "post-DFT" methods, see D. Alfé and M. J. Gillan, *J. Phys.: Condens. Matter* **18** (2006), L435.

For an example of the challenges associated with considering molecules with many degrees of freedom on a surface, see K. Mae and Y. Morikawa, *Surf. Sci.* **553** (2004), L63.

Dipole corrections to DFT calculations for surface calculations are discussed in J. Neugebauer and M. Scheffler, *Phys. Rev. B* **46** (1992), 16067.

APPENDIX   CALCULATION DETAILS

All calculations in this chapter used the PBE GGA functional. For calculations related to Cu surfaces, a cutoff energy of 380 eV and the Methfessel–Paxton scheme was used with a smearing width of 0.1 eV. For calculations related to Si surfaces, the cutoff energy was 380 eV and Gaussian smearing with a width of 0.1 eV was used. The k points were placed in reciprocal space using the Monkhorst–Pack scheme. For all surface calculations, the supercell dimensions in the plane of the surface were defined using the DFT-optimized bulk lattice parameter.

Section 4.5   Surface relaxations were examined using asymmetric slab models of five, six, seven, or eight layers with the atoms in the two bottom layers fixed at bulk positions and all remaining atoms allowed to relax. For Cu(100), the supercell had c(2 × 2) surface symmetry, containing 2 atoms per layer. For Cu(111), $(\sqrt{3} \times \sqrt{3})$R30° surface unit cell with 3 atoms per layer was used. All slab models included a minimum of 23 Å of vacuum along the direction of the surface normal. A $6 \times 6 \times 1$ k-point mesh was used for all calculations.

Section 4.6   The calculations of the surface energy were as in Section 4.5, with the addition of the energy calculation of Cu in the bulk phase. The bulk calculation was carried out on an fcc primitive cell with the DFT-optimized lattice parameter using $11 \times 11 \times 11$ k points.

Section 4.8   The reconstructed Si(001) surface was modeled using a c(4 × 4) asymmetric slab with four layers. The bottom two layers were fixed at bulk positions. Reciprocal space was sampled with $11 \times 11 \times 11$ k points and the vacuum spacing was 17 Å.

Section 4.9   The Cu(100) c(4 × 4)-H overlayer was examined using a four-layer asymmetric slab with the bottom two layers fixed at bulk positions and $6 \times 6 \times 1$ k points.

Section 4.10   The calculations of H adsorption on Cu(100) as a function of coverage were carried out on four-layer slabs of c(1 × 1), c(2 × 2), and c(4 × 4) symmetry using k point meshes of $16 \times 16 \times 1$, $12 \times 12 \times 1$, and $6 \times 6 \times 1$, respectively. The calculation of H_2 was carried out in a box of edge length 20 Å. The optimized H_2 bond length was 0.75 Å.

5

DFT CALCULATIONS OF VIBRATIONAL FREQUENCIES

In the previous chapters, you have learned how to use DFT calculations to optimize the structures of molecules, bulk solids, and surfaces. In many ways these calculations are very satisfying since they can predict the properties of a wide variety of interesting materials. But everything you have seen so far also substantiates a common criticism that is directed toward DFT calculations: namely that it is a *zero temperature* approach. What is meant by this is that the calculations tell us about the properties of a material in which the atoms are localized at "equilibrium" or "minimum energy" positions. In classical mechanics, this corresponds to a description of a material at 0 K.* The implication of this criticism is that it may be interesting to know about how materials would appear at 0 K, but real life happens at finite temperatures.

Much of the following chapters aim to show you how DFT calculations can give useful information about materials at nonzero temperatures. As a starting point, imagine a material that is cooled to 0 K. In the context of classical mechanics, the atoms in the material will relax to minimize the energy of the material. We will refer to the coordinates of the atoms in this state as the equilibrium positions. One of the simplest things that happens (again from

---

*This is true in classical mechanics, but in quantum mechanics it is more accurate to say that atoms are localized around their energy minima. We return to this idea in Section 5.4 when we discuss zero-point energies.

Density Functional Theory: A Practical Introduction. By David S. Sholl and Janice A. Steckel
Copyright © 2009 John Wiley & Sons, Inc.

a classical perspective) if the material is raised to a nonzero temperature is that the atoms in the material will vibrate about their equilibrium positions. From a more correct quantum mechanical perspective, the vibrations that are possible around an atom's equilibrium position contribute to the material's energy even at 0 K via zero-point energies. In many instances, these vibrations can be measured experimentally using spectroscopy, so the frequencies of vibrations are often of great interest. In this chapter, we look at how DFT calculations can be used to calculate vibrational frequencies.

5.1 ISOLATED MOLECULES

We will begin with a simple example, the vibrations of an isolated CO molecule. More specifically, we consider the stretching of the chemical bond between the two atoms in the molecule. For convenience, we assume that the bond is oriented along the x direction in space. The bond length is then defined by $b = x_C - x_O$, where x_C and x_O are the positions of the two atoms. A Taylor expansion for the energy of the molecule expanded around the equilibrium bond length, b_0, gives

$$E = E_0 + (b - b_0)\left[\frac{dE}{db}\right]_{b=b_0} + \frac{1}{2}(b - b_0)^2 \left[\frac{d^2E}{db^2}\right]_{b=b_0} + \cdots \quad (5.1)$$

The first derivative term is exactly zero because it is evaluated at the energy minimum. So, for small displacements about the equilibrium bond length, b_0, $E = E_0 + \alpha/2(b - b_0)^2$ where $\alpha = [d^2E/db^2]_{b=b_0}$. This approach, which neglects the higher order terms in the Taylor expansion, is called the *harmonic approximation*.

How do the atoms move within this harmonic approximation? Treating the C nucleus as a classical particle following Newton's law, we have $F_C = ma_C$ with $F_C = -\partial E/\partial x_C$ and $a = d^2x_C/dt^2$. Similar equations may be written for the position of the O nucleus. A little algebra shows that the equation of motion for the overall bond length is

$$\frac{d^2b(t)}{dt^2} = -\alpha\left(\frac{m_C + m_O}{m_C m_O}\right)(b(t) - b_0). \quad (5.2)$$

The solution of this equation is $b(t) = b_0 + a\cos\omega t$ where a is an arbitrary constant and

$$\omega = \sqrt{\alpha\frac{m_C + m_O}{m_C m_O}}.$$

This means that the bond length oscillates with a characteristic vibrational frequency

$$v = \frac{\omega}{2\pi} = \frac{1}{2\pi} \sqrt{\alpha \frac{m_C + m_O}{m_C m_O}}.$$

To calculate the vibrational frequency of CO using DFT, we first have to find the bond length that minimizes the molecule's energy. The only other piece of information we need to calculate is $\alpha = (d^2E/db^2)_{b=b_0}$. Unfortunately, plane-wave DFT calculations do not routinely evaluate an analytical expression for the second derivatives of the energy with respect to atomic positions. However, we can obtain a good estimate of the second derivative using a finite-difference approximation:

$$\left(\frac{d^2E}{db^2}\right)_{b_0} \cong \frac{E(b_0 + \delta b) - 2E(b_0) + E(b_0 - \delta b)}{(\delta b)^2}. \qquad (5.3)$$

This expression becomes exact when $\delta b \rightarrow 0$, as you can verify by expanding the terms on the right-hand side using the Taylor expansion in Eq. (5.1).

As an example, we can use DFT calculations to evaluate the frequency of the stretching mode for a gas-phase CO molecule using the finite-difference approximation described above. The result from applying Eq. (5.3) for various values of the finite-difference displacement, δb, are listed and plotted in Table 5.1 and Fig. 5.1. For a range of displacements, say from $\delta b = 0.005 - 0.04$ Å, the calculated vibrational frequency is relatively insensitive to the value of δb. For larger and smaller displacements, however, this situation changes markedly. It is important to understand the origins of this behavior. The inaccuracy of the

TABLE 5.1 Vibrational Frequencies of Gas-Phase CO Computed Using DFT for Several Different Finite-Difference Displacements

δb (Å)	Vibrational Frequency (cm^{-1})
0.001	2078
0.005	2126
0.010	2123
0.020	2122
0.040	2132
0.060	2148
0.100	2197
0.200	2433
0.500	4176

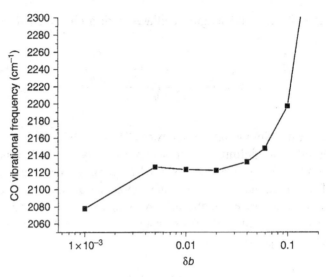

Figure 5.1 Vibrational frequencies of gas-phase CO computed using DFT as a function of finite-difference displacement, δb, in angstroms.

calculation with large δb follows directly from the derivation of Eq. (5.3). This expression is only completely accurate when $\delta b \rightarrow 0$, so it is not surprising that the results are inaccurate if δb is too large.

The inaccuracy of the vibrational frequency in Table 5.1 when $\delta b = 0.001$ Å may seem more surprising since making δb smaller moves Eq. (5.3) toward the limit where it is exact. This expression can only be exact, however, if we can evaluate the energies it uses exactly. It is useful to rewrite Eq. (5.3) as

$$\left(\frac{d^2 E}{db^2}\right)_{b_0} \cong \frac{\Delta E(\delta b) + \Delta E(-\delta b)}{(\delta b)^2}, \tag{5.4}$$

where $\Delta E(x) = E(b_0 + x) - E(b_0)$. For very small values of δb, the energy differences in Eq. (5.4) become too small to accurately evaluate with a DFT calculation. For a CO molecule, changing the bond length by 0.001 Å from its equilibrium position only changes the energy of the molecule by roughly 0.00002 eV. If you look back to the lengthy discussion of convergence in Chapter 3, you will see that performing a calculation that accurately measures this energy difference would be very challenging.

The example of a CO molecule has shown that the displacements used in finite-difference calculations for vibrational frequencies should be not too small and not too large. In practice, it is a good idea to choose displacements that result in energy differences on the order of 0.01–0.10 eV since these energy differences can be calculated accurately without requiring extraordinary care. For most materials, these energy differences correspond to finite-difference displacements in the range of $0.03-0.1$ Å. As with any DFT calculation, it is wise to perform a series of calculations testing the consistency of your numerical results as a function of δb when you begin to examine the vibrational properties of a new material.

Now that we understand how to get a well-converged vibrational frequency for CO from DFT, we can compare this result to experimental data. Experimentally, the stretching frequency of gas-phase CO is 2143 cm^{-1}. This value is ~ 20 cm^{-1} higher than our DFT result. This result is fairly typical of a wide range of vibrational frequencies calculated with DFT. The discrepancy between the DFT result and the true vibrational frequency arises in part because of our harmonic treatment of the vibrations, but is also due to the inexact nature of DFT in solving the Schrödinger equation. We return to this issue in the context of a more general discussion of the accuracy of DFT in Chapter 10.

5.2 VIBRATIONS OF A COLLECTION OF ATOMS

Our calculations in Section 5.1 only looked at the simplest example of a vibration, namely the vibration of an isolated diatomic molecule. How do these vibrations change if the molecule is interacting with other atoms? For example, we could imagine that a CO molecule is chemically bonded to a metal surface. To describe this situation, it is useful to extend the derivation of the vibrational frequency given above to a more general situation where a collection of multiple atoms can vibrate.

We begin by defining a set of N atoms and writing their Cartesian coordinates as a single vector with $3N$ components, $\mathbf{r} = (r_1, \ldots, r_{3N})$. If locating the atoms at $\mathbf{r}_0$ is a local minimum in the energy of the atoms, then it is convenient to define new coordinates $\mathbf{x} = \mathbf{r} - \mathbf{r}_0$. The Taylor expansion of the atom's energy about the minimum at $\mathbf{r}_0$ is, to second order,

$$E = E_0 + \frac{1}{2} \sum_{i=1}^{3N} \sum_{j=1}^{3N} \left[\frac{\partial^2 E}{\partial x_i \, \partial x_j} \right]_{\mathbf{x}=0} x_i x_j. \tag{5.5}$$

The terms in this expression involving first derivatives are zero for the same reason that this was true in Eq. (5.1). If we define

$$H_{ij} = \left[\frac{\partial^2 E}{\partial x_i \, \partial x_j} \right]_{\mathbf{x}=0} \tag{5.6}$$

then these derivatives define a $3N \times 3N$ matrix known as the Hessian matrix. Just as we did in one dimension for the CO molecule, we can now derive an equation of motion for the classical dynamics of the atoms by writing down the force associated with the ith coordinate, F_i. This force is related to the acceleration of this coordinate by $F_i = m_i(d^2 x_i/dt^2)$ where the mass of the atom associated with the ith coordinate is m_i. This force is simultaneously related to the energy of the N atoms by $F_i = -\partial E/\partial x_i$. In matrix form, the equations of motion that emerge from this analysis are

$$\frac{d^2 \mathbf{x}}{dt^2} = -\mathbf{A}\mathbf{x}, \tag{5.7}$$

where the elements of the matrix $\mathbf{A}$ are $A_{ij} = H_{ij}/m_i$. This matrix is called the mass-weighted Hessian matrix.

The equations of motion we just derived have a set of special solutions that are associated with the eigenvectors of $\mathbf{A}$. The eigenvectors of this matrix and their eigenvalues λ are the vectors $\mathbf{e}$ that satisfy $\mathbf{A}\mathbf{e} = \lambda\mathbf{e}$. If we consider the motion of all the atoms starting from the initial condition $\mathbf{x}(t = 0) = a\mathbf{e}$ (for an arbitrary constant a) and $d\mathbf{x}(t = 0)/dt = 0$, then the positions of the atoms have a simple form: $\mathbf{x}(t) = a\cos(\omega t)\mathbf{e}$ with $\omega = \sqrt{\lambda}$. That is, if the initial displacements of the atoms (relative to the energy minimum) point along an eigenvector of $\mathbf{A}$, then the displacements will point along the same eigenvector for all time and the amplitude varies sinusoidally with a frequency defined via the eigenvalue. These special solutions of the equations of motion are called *normal modes*.

In general, the mass-weighted Hessian matrix $\mathbf{A}$ has $3N$ eigenvectors, $\mathbf{e}_1, \ldots, \mathbf{e}_{3N}$. More importantly, the general solution of the equations of motion, Eq. (5.7), is a linear combination of the normal modes:

$$\mathbf{x}(t) = \sum_{i=1}^{3N} [a_i \cos(\omega t) + b_i \sin(\omega t)]\mathbf{e}_i. \tag{5.8}$$

Here the a_i and b_i are a collection of constants that are uniquely determined by the initial positions and velocities of the atoms. This means that the normal modes are not just special solutions to the equations of motion; instead, they offer a useful way to characterize the motion of the atoms for *all* possible initial conditions. If a complete list of the normal modes is available, it can be viewed as a complete description of the vibrations of the atoms being considered.

To determine the full set of normal modes in a DFT calculation, the main task is to calculate the elements of the Hessian matrix. Just as we did for CO in one dimension, the second derivatives that appear in the Hessian matrix can be estimated using finite-difference approximations. For example,

$$H_{ij} = \left(\frac{\partial^2 E}{\partial x_i \, \partial x_j}\right)_{\mathbf{x}=0} \cong \frac{E(\delta x_i, \delta x_j) - 2E_0 + E(-\delta x_i, -\delta x_j)}{\delta x_i \, \delta x_j}. \tag{5.9}$$

This is written using the shorthand notation that $E(\delta x_i, \delta x_j)$ is the energy of the atoms when only the coordinates that are specified are nonzero. Once all the elements of the Hessian matrix are calculated, they are used to define the mass-weighted Hessian matrix, $\mathbf{A}$. The $3N$ normal mode frequencies are then given by the $3N$ eigenvalues of this matrix, $v_i = \sqrt{\lambda_i}/2\pi$. The specific atomic motions giving rise to a particular vibrational frequency are defined by the eigenvector corresponding to that frequency.

Based on the discussion above, you may predict that a frequency calculation on the gas-phase CO molecule should result in $3N = 6$ eigenvalues since an isolated CO molecule has 6 degrees of freedom (three spatial coordinates for each atom). This conclusion is correct, and the five modes that we did not consider before can be determined by analyzing the full Hessian matrix for an isolated molecule. When this is done, three modes defined by rigid-body translation are found. You can think of these as rigid translations of the molecule along the x, y, and z axes that do not change the molecule's bond length. Moving the molecule in this way does not change the molecule's energy, so the resulting eigenvalue of the Hessian is (in theory) zero. Because two independent angles are needed to define a direction in spherical coordinates, there are two independent rotation modes. Again, the eigenvalues associated with these modes are (in theory) zero. The existence of three translational modes and two rotational modes is not a special property of the diatomic molecule we have been considering; any isolated collection of two or more atoms in which all the atoms are allowed to move has these five normal modes.

Because the Hessian matrix is calculated in practice using finite-difference approximations, the eigenvalues corresponding to the translational and rotational modes we have just described are not exactly zero when calculated with DFT. The normal modes for a CO molecule with a finite difference displacements of $\delta b = 0.04$ Å are listed in Table 5.2. This table lists the calculated frequency of each mode and the eigenvector associated with each mode. The CO molecule was aligned along the z axis for this calculation.

We can identify mode 1 in Table 5.2 as the CO stretch by examining the eigenvector to see that the C and O molecules are moving in opposite directions from one another. Similarly, the eigenvectors show that modes 2 and 3 are associated with rotation in the x–z and y–z planes, respectively. The frequencies of these two modes are not found to be zero, instead they are ~ 50–60 cm^{-1}. Modes 4, 5, and 6 are associated with translations in the z, y, and x directions, respectively. Like the rotational modes, the calculated frequencies of these modes are not zero. In fact, each mode is calculated to have an imaginary frequency of small magnitude. The deviation between the calculated frequency for modes 2–5 and the correct theoretical result (namely, frequencies that are exactly zero) occurs because of the small numerical inaccuracies that inevitably occur in calculating a Hessian via finite-difference expressions and a calculation method with finite numerical

TABLE 5.2   DFT-calculated Normal Modes for Gas-Phase CO[a]

Normal Mode	Frequency (cm^{-1})	Atom	Eigenvector (Å) x	Y	Z	Mode Type
1	2132	C	0.00	0.00	**0.76**	CO stretch
		O	0.00	0.00	**−0.65**	
2	59	C	**−0.65**	0.00	0.00	Rotation
		O	**0.73**	0.00	0.00	
3	56	C	0.00	**−0.69**	0.00	Rotation
		O	0.00	**0.70**	0.00	
4	0.10i	C	0.00	0.00	**−0.65**	Translation
		O	0.00	0.00	**−0.76**	
5	4.5i	C	0.00	**0.71**	0.00	Translation
		O	0.00	**0.70**	0.00	
6	7.9i	C	**0.75**	0.00	0.00	Translation
		O	**0.66**	0.00	0.00	

[a]The components of the eigenvectors are listed in Å rounded to two significant figures, with nonzero values in bold for emphasis. The frequencies of modes 4, 5, and 6 are imaginary.

accuracy. Because this outcome is a generic result for plane-wave DFT calculations (or any other flavor of quantum chemistry that uses finite differences to calculate normal modes), it is always important to understand how many zero-frequency modes must exist for any particular collection of atoms. Once this number is known, it is typically easy in a calculated set of normal modes to identify these modes.

5.3   MOLECULES ON SURFACES

Now that we have reviewed the calculation of vibrational frequencies for collections of atoms, let us move on to something more interesting: namely, calculating the vibrational frequency of a CO molecule adsorbed on a surface. As a simple example, we will look at CO adsorbed on the Cu(001) surface. Experimentally, it is known that CO forms a well-ordered layer on this surface with a c(2 × 2) structure (see Fig. 4.17 for a picture of this structure).[1] In this structure, the ratio of CO molecules to metal atoms on the surface is 1 : 2. DFT calculations very similar to those we discussed in Chapter 4 show that the top site is the preferred adsorption site for this ordered structure, with the C atom bonding to a Cu atom and the CO molecule oriented along the surface normal. The calculated CO bond length for the adsorbed CO molecule is 1.158 Å, which is 0.015 Å longer than the (calculated) gas-phase bond length. This lengthening of the equilibrium bond length is reflected in the vibrational frequency of the molecule.

How does the change in the molecule's geometry between the gas phase and the surface affect the vibrations of the molecule? If you guessed that the vibrational frequency of CO might decrease on the surface, you were right. Adsorption of CO on the surface lengthens the CO bond and the longer (weaker) bond will vibrate at a lower frequency than its gas-phase counterpart. This effect can be seen in detail using the normal modes calculated using DFT for a CO molecule on Cu(001), which are listed in Table 5.2. In the calculation of the Hessian matrix, only the degrees of freedom of the adsorbed molecule were included. That is, the positions of the metal atoms in the surface were fixed in their equilibrium positions during all of the finite-difference calculations. The main result in this table is that the CO stretching frequency on the surface, calculated to be 2069 cm^{-1}, has a considerably lower frequency than the gas-phase value, 2132 cm^{-1}. This observation is in good agreement with experiment, where it is found that on the surface the CO stretching frequency is 2086 cm^{-1}, 57 cm^{-1} lower in frequency than the experimentally observed gas-phase frequency.

In Section 5.2, we described how translations and rotations of a gas-phase molecule do not cause changes in the energy of the molecule and therefore, in theory, lead to normal modes with zero frequencies. This description changes when the molecule is on the surface because if we translate the molecule (but not the surface) then the molecule's energy changes because of its interactions with the surface. This behavior can be seen in modes 2–6 in Table 5.3, which all have significant nonzero frequencies. As in the gas-phase results, the largest eigenvalue is for the eigenvector corresponding to the CO stretch mode

TABLE 5.3   Similar to Table 5.2 but for a CO Molecule Adsorbed on Cu(001) in the c(2 × 2) Structure[a]

| Normal Mode | Frequency (cm^{-1}) | Atom | Eigenvector (Å) | | | Mode Type |
			X	Y	z	
1	2069	C	0.00	0.00	**0.77**	CO stretch
		O	0.00	0.00	**−0.64**	
2	322	C	0.00	0.00	**0.64**	Translation
		O	0.00	0.00	**0.77**	
3	279	C	**0.26**	**0.87**	0.00	Rotation
		O	**−0.12**	**−0.40**	0.00	
4	279	C	**−0.87**	**0.26**	0.00	Rotation
		O	**0.40**	**−0.12**	0.00	
5	46	C	**0.08**	**0.41**	0.00	Translation
		O	**0.18**	**0.89**	0.00	
6	38	C	**−0.41**	**0.08**	0.00	Translation
		O	**−0.89**	**0.18**	0.00	

[a]The CO molecule is aligned in the z direction; nonzero values in bold for emphasis.

(mode 1). However, the next-to-largest eigenvalue is now far from zero; it is $322 \, cm^{-1}$. This mode is associated with translation of the CO molecule in the z direction. If you think about this for a moment, you can see that this mode is better described as the stretching of the bond between the carbon atom and the surface copper atom. The other four modes listed in Table 5.3 are best described as frustrated rotations and translations, although these labels are descriptive rather than precise. Rotation in the $x-z$ and $y-z$ planes now correspond to bending of the Cu–C–O angle. These are modes 3 and 4 in the Table 5.3. Modes 5 and 6 are most closely associated with translations in the y and x directions. Because of the symmetry of this example, the frequencies of modes 3 and 4 are equal, as are the frequencies of modes 5 and 6 (within numerical accuracy).

5.4   ZERO-POINT ENERGIES

Our discussion of vibrations has all been within the context of the harmonic approximation. When using this approach, each vibrational mode can be thought of as being defined by a harmonic oscillator. The potential energy of a one-dimensional harmonic oscillator is

$$E = E_0 + \frac{k}{2}x^2, \tag{5.10}$$

where x is a coordinate chosen so that $x = 0$ is the equilibrium position. The behavior of a harmonic oscillator within classical mechanics is quite simple, and we have used this behavior in the previous two sections to understand vibrations. A more precise description, however, would use quantum mechanics to describe this situation. Fortunately, the harmonic oscillator is one of the small number of quantum mechanical problems that can be completely solved in a simple form. The details of these solutions are available in any textbook on quantum mechanics—several are listed at the end of the chapter. Here, we focus on one specific quantum mechanical phenomenon that is relevant to DFT calculations, namely, zero-point energy.

What is the lowest possible energy for the harmonic oscillator defined in Eq. (5.10)? Using classical mechanics, the answer is quite simple; it is the equilibrium state with $x = 0$, zero kinetic energy and potential energy E_0. The quantum mechanical answer cannot be quite so simple because of the Heisenberg uncertainty principle, which says (roughly) that the position and momentum of a particle cannot both be known with arbitrary precision. Because the classical minimum energy state specifies both the momentum and position of the oscillator exactly (as zero), it is not a valid quantum mechanical state. The lowest quantum mechanical energy that can exist for the harmonic oscillator is

$$E = E_0 + \frac{h\nu}{2}, \tag{5.11}$$

where h is Planck's constant and ν is the (classical) vibrational frequency of the oscillator. The difference between this energy and the classical minimum energy is called the *zero-point energy*. The other energy levels that exist for a quantum mechanical harmonic oscillator are $E_n = E_0 + (n + \frac{1}{2})h\nu$ for positive integers n.

If we now think about a set of atoms that define a set of normal modes, we can determine the zero-point energy of each mode independently. The minimum energy that can be achieved by the set of atoms is then

$$E = E_0 + \sum_i \frac{h\nu_i}{2}. \tag{5.12}$$

Here, E_0 is the energy that we would obtain from a DFT calculation and ν_i are the normal mode frequencies. At the beginning of this chapter we discussed how one way to view DFT results is as providing information at 0 K, where atoms and molecules adopt their minimum energies. You can now see that to be precise in this description we should include zero-point energies as shown in Eq. (5.12). In practice, this approach is not commonly used. One justification for neglecting zero-point energies is a pragmatic one; computing the normal mode frequencies is much more computationally expensive than just performing a DFT energy minimization!

Fortunately, there is also a good physical justification for not including zero-point energies when reporting energies from DFT calculations. In many circumstances the changes in zero-point energy between states of interest are relatively small, even when the total zero-point energy is considerable. As an example, we can look at the adsorption of CO on Cu(100). If we neglect zero-point energies, the adsorption energy of the molecule is

$$E_{ads} = (E_{CO} + E_{Cu}) - E_{CO/Cu}. \tag{5.13}$$

As discussed in Chapter 4, this expression includes results from three independent DFT calculations: the energy of gas-phase CO, E_{CO}, the energy of the bare Cu surface, E_{Cu}, and the energy of the Cu surface with CO adsorbed upon it, $E_{CO/Cu}$. For the example we introduced above, this gives $E_{ads} = 0.71$ eV. If we include zero-point energies by assuming that only vibrations of the CO molecule contribute, then the adsorption energy is

$$E_{ads} = (E_{CO} + E_{Cu}) - E_{CO/Cu} + \frac{h\nu_{CO}}{2} - \sum_{i=1}^{6} \frac{h\nu_i}{2}. \qquad (5.14)$$

The zero-point energy terms on the right-hand side of this expression come from the one (six) vibrational mode that exists for the gas-phase (adsorbed) molecule. Inclusion of the zero-point energies results in $E_{ads} = 0.65$ eV, or a difference of about 0.06 eV.[†] Compared to the net adsorption energy, this contribution from the zero-point energies is not large. Most of this net zero-point energy comes from the five modes that exist on the surface with nonzero frequencies in Eq. (5.14), not the CO stretch. The zero-point energy correction to the adsorption energy associated just with the change in the CO stretching frequency between the gas phase and the adsorbed molecule is only ~0.004 eV.

As a qualitative rule, zero-point energy effects become more important for problems involving light atoms. Specifically, this means that zero-point energies may be especially important in problems that involve H atoms. As an example, we can compare the energy of an H atom on the surface of Cu(100) with the energy of an H atom inside bulk Cu. There are two kinds of interstitial sites that can be considered inside an fcc metal, as illustrated in Fig. 5.2, the sixfold octahedral site and the fourfold tetrahedral site. We computed the energy of atomic H in one of these interstitial sites relative to the energy of atomic H on the Cu(100) surface using

$$\Delta E = (E_{H/Cu(100)} + E_{bulk}) - (E_{Cu(100)} + E_{H/bulk}), \qquad (5.15)$$

where E_{bulk} and $E_{Cu(100)}$ are the DFT-calculated energies of a supercell containing bulk Cu and a Cu(100) slab, respectively, and $E_{H/bulk}$ and $E_{H/Cu(100)}$ are the energies of the same supercells once an H atom is introduced into the supercell. Using this definition, a negative value of ΔE implies that the interstitial site being considered is more energetically favorable than the surface site.

The results from these calculations are summarized in Table 5.4. The results without including zero-point energies indicate that the bulk octahedral site is slightly more energetically preferred than the surface site, but that the bulk tetrahedral site is less favorable. The calculated zero-point energies for the surface site and the bulk octahedral site are similar in magnitude, so the relative

---

[†]Although the change in this energy from the zero-point energies is arguably not large, there are situations where it could be unwise to neglect it. This would be the case, for example, if you were attempting to decide which of several possible adsorption sites was most energetically favorable and the sites differed in (classical) energy by an amount similar to the zero-point energy corrections.

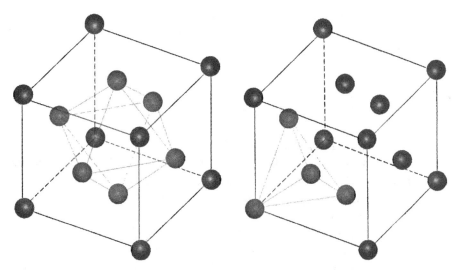

Figure 5.2 Schematic views of the octahedral (*left*) and tetrahedral (*right*) interstitial sites that exist inside an fcc metal. The octahedral site is formed by connecting six face center atoms, while the tetrahedral site is formed by connecting four adjacent nearest-neighbor atoms.

energy of these two sites only changes slightly once zero-point energies are included. The zero-point energy of atomic H in the bulk tetrahedral site is considerably larger than for the other two sites. Physically, this means that the H atom is more confined in the tetrahedral site than it is in the octahedral site, giving the tetrahedral H atom higher vibrational frequencies. If we were interested in describing the energy difference between the tetrahedral site and the other two sites, it is clear that it is likely to be important that zero-point energies are included in this comparison.

A final comment related to zero-point energies is that all of these results above relate to the lowest energy state that a set of atoms can take at 0 K. When thinking about the same set of atoms at a nonzero temperature, it is often useful to know the average energy at the specified temperature. If we continue using the harmonic approximation to describe each normal mode, then this average energy summed over all energy levels is one of the few

TABLE 5.4 Relative Energies of H on Cu(100) and of H in Interstitial Sites in Bulk Cu as Defined in Eq. (5.15) (all energies in eV)

Site	Relative Energy (no ZPE)	ZPE	Relative Energy (with ZPE)
Cu(100) surface	0	+0.12	0
Bulk octahedral	−0.05	+0.14	−0.03
Bulk tetrahedral	+0.12	+0.21	+0.21

results from quantum statistical mechanics that can be written in a simple, closed form. For each harmonic oscillator defined by Eq. (5.10), the average energy at temperature T is

$$\overline{E}(T) = \frac{h\nu}{2} \frac{[\exp(\beta h\nu/2) + \exp(-\beta h\nu/2)]}{[\exp(\beta h\nu/2) - \exp(-\beta h\nu/2)]}, \tag{5.16}$$

where $\beta = 1/k_B T$. This expression reduces to Eq. (5.11) as $T \to 0$, so it is consistent with the 0 K results we have already discussed. In the opposite limit, $\overline{E}(T) \to kT$ for very high temperatures; this is the classical limit, where the average energy of the harmonic oscillator is independent of its vibrational frequency.

An implication of Eq. (5.16) is that when zero-point energies are considered, the energy difference between two configurations for a set of atoms is temperature dependent.[‡] In the limit of low temperatures, the energy difference is given by including the zero-point energy of each configuration. At sufficiently high temperatures, each normal mode reaches the classical limit,

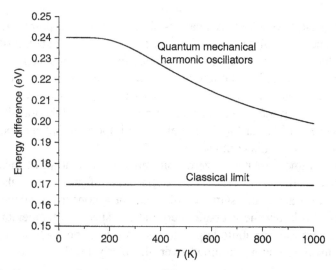

Figure 5.3 Temperature-dependent energy difference between an H atom in an O site and a T site in bulk Cu calculated by applying Eq. (5.16) to each normal mode of the H atom. The low-temperature limit and the classical limit for this energy difference can be determined from Table 5.4.

[‡]This is a consequence of describing the harmonic oscillators with quantum mechanics. In a purely classical description, the average energy for each harmonic oscillator is simply $k_B T$, so the energy difference between two configurations is independent of T.

so the energy difference between two configurations is just the difference between the classical energies (i.e., no zero-point energies are needed). This idea is illustrated in Fig. 5.3, which shows the T-dependent energy difference between having an H atom in the O and T sites inside bulk Cu. At temperatures below $\sim$200 K, this energy difference is the zero-point energy corrected value from Table 5.4, that is, 0.24 eV. At higher temperatures, the energy difference decreases slightly, smoothly approaching the classical limit (0.17 eV) at extremely high temperatures.

5.5 PHONONS AND DELOCALIZED MODES

Every example of a vibration we have introduced so far has dealt with a localized set of atoms, either as a gas-phase molecule or a molecule adsorbed on a surface. Hopefully, you have come to appreciate from the earlier chapters that one of the strengths of plane-wave DFT calculations is that they apply in a natural way to spatially extended materials such as bulk solids. The vibrational states that characterize bulk materials are called phonons. Like the normal modes of localized systems, phonons can be thought of as special solutions to the classical description of a vibrating set of atoms that can be used in linear combinations with other phonons to describe the vibrations resulting from any possible initial state of the atoms. Unlike normal modes in molecules, phonons are spatially delocalized and involve simultaneous vibrations in an infinite collection of atoms with well-defined spatial periodicity. While a molecule's normal modes are defined by a discrete set of vibrations, the phonons of a material are defined by a continuous spectrum of phonons with a continuous range of frequencies. A central quantity of interest when describing phonons is the number of phonons with a specified vibrational frequency, that is, the vibrational density of states. Just as molecular vibrations play a central role in describing molecular structure and properties, the phonon density of states is central to many physical properties of solids. This topic is covered in essentially all textbooks on solid-state physics—some of which are listed at the end of the chapter.

Using DFT calculations to predict a phonon density of states is conceptually similar to the process of finding localized normal modes. In these calculations, small displacements of atoms around their equilibrium positions are used to define finite-difference approximations to the Hessian matrix for the system of interest, just as in Eq. (5.3). The mathematics involved in transforming this information into the phonon density of states is well defined, but somewhat more complicated than the results we presented in Section 5.2. Unfortunately, this process is not yet available as a routine option in the most widely available DFT packages (although these calculations are widely used within some sections of the research community). Readers interested in exploring this topic further can consult the Further Reading section at the end of the chapter.

EXERCISES

1. Use DFT calculations to determine the vibrational frequency of gas-phase N_2. Compare your result with experimental data. How do your results depend on the displacement used in the finite-difference approximation?

2. Use DFT calculations to determine the vibrational frequencies of gas-phase ammonia, NH_3. Compare your results with experimental data and interpret each frequency in terms of the type of vibration associated with it.

3. Hydrogen atoms on $Cu(111)$ can bind in two distinct threefold sites, the fcc sites and hcp sites. Use DFT calculations to calculate the classical energy difference between these two sites. Then calculate the vibrational frequencies of H in each site by assuming that the normal modes of the adsorbed H atom. How does the energy difference between the sites change once zero-point energies are included?

4. In the previous exercise you assumed that the H vibrations were uncoupled from the metal atoms in the surface. A crude way to check the accuracy of this assumption is to find the normal modes when the H atom and the three metal atoms closest to it are all allowed to move, a calculation that involves 12 normal modes. Perform this calculation on one of the threefold sites and identify which 3 modes are dominated by motion of the H atom. How different are the frequencies of these modes from the simpler calculation in the previous exercise?

REFERENCE

1. R. Ryberg, Carbon-Monoxide Adsorbed on Cu(100) Studied by Infrared-Spectroscopy, *Surf. Sci.* **114** (1982), 627–641.

FURTHER READING

The quantum mechanical treatment of harmonic oscillators is described in essentially all books on quantum mechanics. Several good examples are:

P. W. Atkins and R. S. Friedman, *Molecular Quantum Mechanics*, Oxford University Press, Oxford, UK, 1997.

D. A. McQuarrie, *Quantum Chemistry*, University Science Books, Mill Valley, CA, 1983.

M. A. Ratner and G. C. Schatz, *Introduction to Quantum Mechanics in Chemistry*, Prentice Hall, Upper Saddle River, NJ, 2001.

J. Simons and J. Nichols, *Quantum Mechanics in Chemistry*, Oxford University Press, New York, 1997.

The concept of phonons describing vibrations in extended solids is a core topic in solid-state physics. Two classic books for this field are:

N. W. Ashcroft and N. D. Mermin, *Solid State Physics*, Saunders College Publishing, Orlando, 1976.

C. Kittel, *Introduction to Solid State Physics*, Wiley, New York, 1976.

For a concise overview of using DFT calculations to determine the phonon density of states for bulk materials:

G. J. Ackland, *J. Phys. Condens. Matter* 14 (2002), 2975–3000.

APPENDIX   CALCULATION DETAILS

All calculations in this chapter used the PBE GGA functional and a plane-wave basis set including waves with an energy cutoff of 380 eV. For all surface calculations, the supercell dimensions in the plane of the surface were defined using the DFT-optimized bulk lattice parameter.

Section 5.1   The gas phase calculations of the CO stretching vibration were carried out in a box that was approximately $10 \times 10 \times 10$ Å and the CO molecule was aligned with the z axis. The dimensions of the box were intentionally made to be slightly different, in order to avoid having a perfectly symmetric box. Gaussian smearing was used with a smearing width of 0.1 eV. Reciprocal space was sampled at the Γ-point only. Each ion was displaced by a small positive and a small negative amount in each direction. The finite difference calculations were carried out using a range of values for the atomic displacements as described in the chapter.

Section 5.2   The gas phase calculations of the CO stretching vibrations were as in Section 5.1, except that a finite difference of 0.04 Å was used.

Section 5.3   The vibrations of CO adsorbed on the Cu(100) surface were examined using an asymmetric slab model of 4 layers with the atoms in the two bottommost layers fixed at bulk positions and all remaining atoms allowed to relax previous to the calculation of the normal modes. The supercell had c(2×2) surface symmetry, containing 2 metal atoms per layer. A $6 \times 6 \times 1$ k-point mesh was used for all calculations. The smearing method of Methfessel and Paxton of order 2 was used with a smearing width of 0.1 eV.

Section 5.4 The adsorption of H on Cu(100) was examined using an asymmetric slab model of 4 layers with the atoms in the two bottommost layers fixed at bulk positions and all remaining atoms allowed to relax. The k point mesh was $6 \times 6 \times 1$ and the method of Methfessel and Paxton of order 2 was used with a smearing width of 0.1 eV. The adsorption of H in interstitial sites was examined using a supercell of fcc Cu containing 64 metal atoms supercell. All atoms in the 65 atom supercell were allowed to relax during geometry optimization although the dimensions of the supercell were held fixed at bulk-optimized values for pure Cu. Vibrational frequencies for interstitial H were determined using finite differences as discussed in the chapter and used two displacements per H atom of 0.04 Å each.

6

CALCULATING RATES OF CHEMICAL PROCESSES USING TRANSITION STATE THEORY

There are a vast number of situations where the rates of chemical reactions are of technological or scientific importance. In Chapter 1 we briefly considered the use of heterogeneous catalysts in the synthesis of ammonia, NH_3, from N_2 and H_2. Perhaps the most fundamental property of a good catalyst for this process is that it greatly increases the net reaction rate when compared to other possible catalysts. In a different realm, the atmospheric chemistry that led to the formation of an ozone hole over the Antarctic included a complicated series of chemical reactions that took place on ice crystals in polar stratospheric clouds. Understanding the rates of these reactions was crucial in developing a mechanistic description of atmospheric ozone depletion. To give one more example, radiation damage of solid materials often results in the formation of localized defects such as lattice vacancies. To predict the long-term impact of these defects on material stability, the rate at which defects can move within a material must be known (among many other things).

You now have enough experience with DFT calculations to imagine how calculations could be performed that would be relevant for each of the three examples listed above. For instance, DFT calculations could be used to determine the relative energy of various kinds of lattice defects that could potentially exist in a solid material. Similar calculations could be used to determine the equilibrium positions of reactive molecules on the surfaces of ice crystals that could be thought of as mimics for polar stratospheric clouds.

Density Functional Theory: A Practical Introduction. By David S. Sholl and Janice A. Steckel
Copyright © 2009 John Wiley & Sons, Inc.

But these calculations only give information on minimum energy states. In the language introduced in Chapter 5, these calculations give only 0 K information. This is a bit too cold even for polar stratospheric clouds!

In this chapter we take up the question of how DFT calculations can be used to calculate the rates of chemical processes. We have deliberately used the word "processes" to include not just chemical reactions in which the chemical compounds at the end of the process are different from the starting point of the process (i.e., chemical reactions) but also situations such as the motion of a defect inside a solid where the starting and ending points can be thought of as chemically identical. In both cases, a useful way to think about these processes is in terms of the overall energy surface defined by a set of N atoms, $E(\mathbf{R}_1, \ldots, \mathbf{R}_N)$. This energy surface has, in general, multiple local minima, $E_1, E_2, E_3, \ldots$. The most fundamental process we can consider is one in which the configuration of the N atoms moves from one minimum, E_i, to another minimum, E_j, without passing through any other minima. Our main task in this chapter is to show how DFT calculations can be used to accurately define the rates of processes of this kind.

6.1   ONE-DIMENSIONAL EXAMPLE

Throughout this chapter we will consider a specific example of a chemical process where we might want to determine the rate of the process: the diffusion of an Ag atom on an Cu(100) surface. This example has the virtue of being easy to visualize and includes many of the characteristics that appear in chemical processes that may have greater physical or technological interest.

Adsorption of an Ag atom on Cu(100) is illustrated schematically in Fig. 6.1. A useful way to visualize the energy of the Ag atom is to fix the

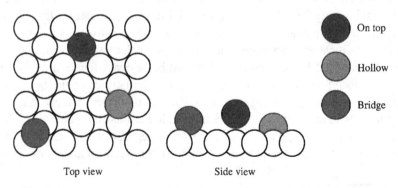

Figure 6.1   Schematic representation of the adsorption of Ag on the Cu(100) surface showing an Ag atom adsorbed at the on-top, fourfold hollow, and bridge sites on the Cu(100) surface.

position of this atom in the plane of the surface (the $x-y$ plane) and to then minimize the energy while allowing the Ag atom to move in the direction of the surface normal (the z direction) and all the surface atoms to relax. This procedure gives us a two-dimensional energy surface, $E(x, y)$, that defines the minimum energy of the Ag atom at any point on the surface. This surface is illustrated in Fig. 6.2. There are three types of critical points on this surface that correspond to the high-symmetry surface sites shown in Fig. 6.1. The fourfold sites are the only energy minima on the surface. The bridge sites are first-order saddle points on the energy surface; moving from a bridge site toward a fourfold site lowers the Ag atom's energy, but moving from a bridge site toward an on-top site increases the energy. The on-top sites are second-order saddle points because moving in either direction in the plane of the surface reduces the Ag atom's energy whereas moving up or down the surface normal will increase this energy.

There are an infinite number of trajectories that an Ag atom could follow in moving from one hollow site to another hollow site, but one that plays a special role in understanding the rates of these transitions is the path along which the change in energy during motion between the sites is minimized. From

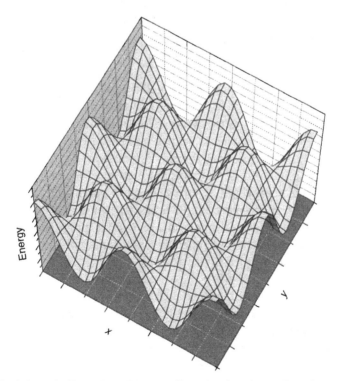

Figure 6.2 Schematic illustration of the two-dimensional energy surface, $E(x, y)$, of an Ag atom on Cu(100). The local minima are the fourfold surface sites.

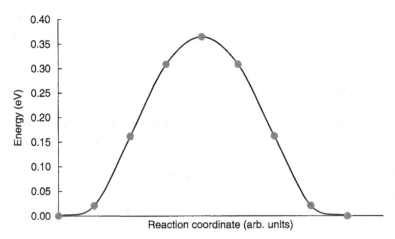

Figure 6.3   DFT-calculated energies for Ag moving along the minimum energy path between two fourfold sites on Cu(100). Energies are relative to the energy of Ag in the fourfold hollow site. The reaction coordinate is a straight line in the $x-y$ plane connecting two adjacent energy minima.

Fig. 6.2, we can see that this path will pass through the saddle point that exists at the bridge site between the two local minima. This special path is called the *minimum energy path* for the process. Figure 6.3 shows the energy of an Ag atom as it is moved between two fourfold sites along the minimum energy path for motion on Cu(100). To appreciate this result, it is useful to compare the range of energies that must be explored by the Ag atom with the typical energies that are available to a single atom at temperature T. A central result from statistical mechanics is that when a set of atoms is in equilibrium at a temperature T, the average energy available to each degree of freedom in the system is $k_B T/2$, where k_B is Boltzmann's constant. At room temperature, $k_B T/2$ is 0.013 eV. From the calculations used to produce Fig. 6.3, we can determine that the energy barrier for an Ag atom moving from a hollow to a bridge site is 0.36 eV, more than 30 times the amount of average thermal energy. It is reasonable to conclude from Fig. 6.3 that an Ag atom with a "typical" amount of energy at room temperature will usually be found very close to the energy minimum defined by the fourfold site.

The energy curve shown in Fig. 6.3 characterizes the energy only in terms of the position of the Ag atom. On a real metal surface, energy will continually be exchanged between the atoms in the system as they collide, and each atom will move about as dictated by the instantaneous forces it feels due to all the other atoms. This means that the Ag atom will from time to time gain considerably more energy than the typical thermal energy, allowing it the possibility of crossing the energy barrier shown in Fig. 6.3. In situations where the energy required for a chemical process is considerably larger than the typical thermal

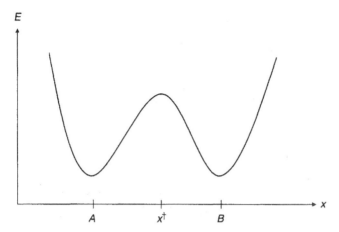

Figure 6.4 One-dimensional energy profile showing two local minima separated by a transition state.

energy, the net rate of the process can be calculated using the transition state theory (TST).

Figure 6.4 shows a schematic view of a one-dimensional energy profile that has the same shape as the result in Fig. 6.3. Figure 6.4 is drawn in terms of a reaction coordinate, x, that measures distance along the path connecting the two local minima at $x = A$ and $x = B$. The saddle point located at $x^\dagger$ that separates the two minima is called the *transition state*. We will refer to all points in Fig. 6.4 to the left (right) of the transition state as state A (state B).

Transition state theory gives the rate of hopping from state A to state B as

$$k_{A \to B} = \left(\tfrac{1}{2}\right) \times (\text{average thermal velocity of hopping atom})$$
$$\times \left(\text{probability of finding atom at } x = x^\dagger\right) \qquad (6.1)$$

The factor of one-half that appears here is because of all the possible ways that atoms could appear at the transition state, only half of them have velocities that correspond to movement to the right (the other half are moving to the left). The probability that is mentioned in the last term in this expression is defined relative to all possible positions of the atom in state A when the material is at thermodynamic equilibrium at temperature T. In this situation, the probability of observing the atom at any particular position is

$$p(x) \propto \exp\left[-\frac{E(x)}{k_B T}\right]. \qquad (6.2)$$

The factor of $1/k_B T$ shows up so frequently that it is given its own symbol: $\beta = 1/k_B T$.

Using this notation,

$$\text{Probability of finding atom at } x = x^\dagger = \frac{e^{-\beta E(x^\dagger)}}{\int_A dx\, e^{-\beta E(x)}}, \qquad (6.3)$$

where the integral is taken over all positions within the minimum associated with $x = A$. The velocities of individual atoms within a material in thermodynamic equilibrium at temperature T follow the so-called Maxwell–Boltzmann distribution. The average velocity of an atom from this distribution can be calculated exactly:

$$\text{Average thermal velocity of hopping atom} = \sqrt{\frac{2}{\beta \pi m}}, \qquad (6.4)$$

where m is the atom's mass. Putting all of this together,

$$k_{A \to B} = \frac{1}{2} \sqrt{\frac{2}{\beta \pi m}} \frac{e^{-\beta E^\dagger}}{\int_A dx\, e^{-\beta E(x)}}. \qquad (6.5)$$

Here, we have defined $E^\dagger = E(x^\dagger)$.

The equation above is not especially helpful because of the integral that appears in the denominator. This situation can be simplified by remembering that an atom with typical thermal energy near the minimum at $x = A$ makes only small excursions away from the minimum. We can then use an idea from our analysis of vibrations in Chapter 5 and expand the energy near the minimum using a Taylor series. That is, we approximate the energy as

$$E(x) \cong E_A + \frac{k}{2}(x - x_A)^2. \qquad (6.6)$$

We showed in Chapter 5 that k is related to the vibrational frequency of the atom in the potential minimum by $v = \frac{1}{2\pi} \sqrt{\frac{k}{m}}$. This means that

$$\int_A dx\, e^{-\beta E(x)} \cong e^{-\beta E_A} \int_A dx\, e^{-\beta k(x - x_A)^2/2}. \qquad (6.7)$$

Because the function inside the integral on the right-hand side, decreases extremely rapidly as $|x - x_A|$ increases, we have

$$\int_A dx\, e^{-\beta E(x)} \cong e^{-\beta E_A} \int_{-\infty}^{+\infty} dx\, e^{-\beta k(x-x_A)^2/2} = \sqrt{\frac{2\pi}{\beta k}}\, e^{-\beta E_A}. \tag{6.8}$$

This means that the overall rate constant has a very simple form:

$$k_{A\to B} = \nu \exp\left(-\frac{E^\dagger - E_A}{k_B T}\right). \tag{6.9}$$

Because of the way the energy was approximated in Eq. (6.6), this result is called *harmonic transition state theory*. This rate only involves two quantities, both of which are defined in a simple way by the energy surface: ν, the vibrational frequency of the atom in the potential minimum, and $\Delta E = E^\dagger - E_A$, the energy difference between the energy minimum and the transition state. ΔE is known as the activation energy for the process. The overall rate, $k_{A\to B} = \nu \exp(-\Delta E/k_B T)$, is the well-known Arrhenius expression for a chemical rate.*

It is not too hard to roughly estimate the prefactor in Eq. (6.9). A "typical" atomic vibration has a period of 0.1–1 ps, so there are 10^{12}–10^{13} vibrations per second. This means that using $\nu = 10^{12}$–10^{13} s^{-1} in Eq. (6.9) gives a reasonable estimate of this quantity without performing any calculations at all. To compare this to the example we have been considering of an Ag atom on Cu(100), computing the vibrational frequency of the Ag atom along the reaction coordinate between two fourfold sites gives $\nu = 1.94 \times 10^{12}$ s$^{-1} = 1.94$ THz, in good agreement with the simple estimate. Using the calculated activation energy for Ag hopping on Cu(100), 0.36 eV, and the Ag atom's vibrational frequency, Eq. (6.9) can be used to predict the hopping rate of an Ag atom. The resulting rates at several temperatures are listed in Table 6.1. The most important feature of this rate (and the feature that is characteristic of all activated processes) is that it changes by orders of magnitude as the temperature is changed over a relatively small range. Because this property of activated rates is dictated by the activation energy in Eq. (6.9), not the vibrational frequency, it is common to use the simple estimate that $\nu = 10^{12}$–10^{13} s^{-1} rather than to perform calculations to assign a more precise value to ν.

---

*Named after Svante Arrhenius, a Swedish chemist who received the lowest possible passing grade for his Ph.D. thesis, which contained work that later earned him the Nobel prize. In 1896, Arrhenius was the first person to suggest a connection between increasing levels of atmospheric CO_2 and global temperatures.

TABLE 6.1   Rates for Ag Hopping Calculated via Eq. (6.9) on Cu(100)

Temperature (K)	Rate (s^{-1})
150	0.97
300	1.4×10^6
450	1.5×10^8

The role of the vibrational frequency and activation energy on the hopping rate is shown in a different way in Fig. 6.5. The solid lines in this figure show the one-dimensional TST results calculated using three estimates for the vibrational frequency: the frequency calculated via DFT and the two "simple" estimates, 10^{12} and 10^{13} s^{-1}. The dashed lines show the result from using the DFT-calculated frequency but adding or subtracting 0.05 eV to the DFT-calculated activation energy. The variation in the hopping rate caused by the changes in the activation energy are considerably larger than the variations that arise when changing the vibrational frequency. The main message from this figure is that it is important to calculate activation energies as precisely as possible when applying transition state theory, but that calculating vibrational frequencies precisely is less important.

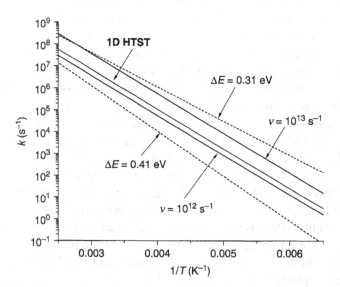

Figure 6.5   Hopping rate for an Ag atom on Cu(100) as predicted with one-dimensional harmonic transition state theory (1D HTST). The other two solid lines show the predicted rate using the DFT-calculated activation energy, $\Delta E = 0.36$ eV, and estimating the TST prefactor as either 10^{12} or 10^{13} s^{-1}. The two dashed lines show the prediction from using the 1D HTST prefactor from DFT ($v = 1.94 \times 10^{12}$ s^{-1}) and varying the DFT-calculated activation energy by ± 0.05 eV.

6.2 MULTIDIMENSIONAL TRANSITION STATE THEORY

The results above for Ag moving on Cu(100) cannot give the whole story for this process because reducing the problem to a one-dimensional minimum energy path is an approximation. If the Ag atom can gain enough energy to get over the transition state on the minimum energy path, then it can also sample locations with similar energies that do not lie on the path. To account for these effects, we need to use a multidimensional version of transition state theory. Fortunately, the net outcome from doing this is almost as easy to interpret as one-dimensional transition state theory.

The generalization of Eq. (6.5) to multiple dimensions is

$$k_{A \to B} = \frac{1}{2} \sqrt{\frac{2}{\beta \pi m}} \frac{\int d\mathbf{r} \, e^{-\beta E(\mathbf{r})} \Big|_{x=x^\dagger}}{\int_A d\mathbf{r} \, e^{-\beta E(\mathbf{r})}}. \tag{6.10}$$

Here, $\mathbf{r}$ is the vector of all relevant particle coordinates and x is one coordinate chosen to define the reaction coordinate. The trick to rewriting this rate in a usable way is to treat both of the integrals that appear in it with the harmonic approximation we used in Section 6.1. For the denominator, we expand the energy surface around the energy minimum at $\mathbf{r} = \mathbf{r}_A$:

$$E = E_A + \frac{1}{2} \sum_{i=1}^{N} \sum_{j=1}^{N} \left[\frac{\partial^2 E}{\partial r_i \, \partial r_j} \right]_{\mathbf{r}_A = 0} (r_i - r_{i,A})(r_j - r_{j,A}). \tag{6.11}$$

We have seen this expression before in Chapter 5, where it was the starting point for describing vibrational modes. We found in Chapter 5 that a natural way to think about this harmonic potential energy surface is to define the normal modes of the system, which have vibrational frequencies v_i ($i = 1, \ldots, N$). The normal modes are the special solutions to the equations of motion that have the form

$$(\mathbf{r} - \mathbf{r}_A) \sim \cos(2\pi v_i t) \mathbf{e}_i, \tag{6.12}$$

where $\mathbf{e}_i$ is the vector defining the direction of the ith normal mode.

We can use exactly the same idea for the integral in the numerator of Eq. (6.10). The Taylor series expansion for the energy expanded around the transition state is, to second order,

$$E = E^\dagger + \frac{1}{2} \sum_{i=1}^{N} \sum_{j=1}^{N} \left[\frac{\partial^2 E}{\partial r_i \, \partial r_j} \right]_{\mathbf{r}=\mathbf{r}^\dagger} \left(r_i - r_i^\dagger \right) \left(r_j - r_j^\dagger \right). \tag{6.13}$$

**TABLE 6.2   Vibrational Frequencies in THz for Ag
Atom in Fourfold Site and Bridge Site on Cu(100)**

Fourfold Site (Energy Minimum)	Bridge Site (Transition State)
3.36	3.60
1.94	2.17
1.89	1.12i

No first derivative terms appear here because the transition state is a critical point on the energy surface; at the transition state all first derivatives are zero. This harmonic approximation to the energy surface can be analyzed as we did in Chapter 5 in terms of normal modes. This involves calculating the mass-weighted Hessian matrix defined by the second derivatives and finding the N eigenvalues of this matrix.

To provide a specific example, consider again an Ag atom hopping out of an fourfold site on Cu(100). We will include only the three coordinates needed to define the position of the Ag atom, so $N = 3$. Using the methods of Chapter 5, we can calculate the eigenvalues of the Hessian matrix with the Ag atom at its energy minimum (the fourfold site on the surface) and at its transition state (the bridge site). In both cases, the coordinates of the surface Cu atoms were relaxed while finding the critical point but were then fixed while calculating the Hessian matrix. To interpret these eigenvalues physically, recall that the vibrational frequency of the normal modes are defined by $v_i = \sqrt{\lambda_i}/2\pi$. The frequencies defined by this expression for the energy minimum and transition state are listed in Table 6.2.

One striking feature of these frequencies is that at the transition state, one of them is imaginary (i.e., the eigenvalue associated with this vibrational mode is negative). If we write this frequency as $v_3 = i\alpha$, then the form of the normal mode solution given in Eq. (6.12) is

$$\left(\mathbf{r} - \mathbf{r}^\dagger\right) \sim \cos(2\pi i \alpha t)\mathbf{e}_3 = \tfrac{1}{2}\left(e^{+2\pi\alpha t} + e^{-2\pi\alpha t}\right)\mathbf{e}_3. \qquad (6.14)$$

This solution means that any displacement away from the transition state along the direction defined by $\mathbf{e}_3$ will grow exponentially with time, unlike the situation for normal modes with real frequencies, which correspond to sinsusoidal oscillations.[†] In more straightforward terms, the transition state is a point on the potential energy surface that is a minimum in all directions but one. At

---

[†]Unlike the oscillating normal mode solutions, Eq. (6.14) is only valid for short times because the trajectory it defines soon moves far enough away from the transition state that the truncated Taylor expansion in Eq. (6.13) is no longer valid.

the energy minimum, all three frequencies in Table 6.2 are real. By symmetry, the two frequencies that correspond to motion in the plane of the surface should be exactly equal. The calculated frequencies in fact differ by 0.05 THz, giving one indication of the level of numerical precision that can be expected in calculations of this kind.

The example above used 3 degrees of freedom, but the main features of this example are present for any transition state. Specifically, a transition state in a system with N degrees of freedom is characterized by $(N - 1)$ real vibrational modes and one imaginary mode, while a local minimum has N real vibrational modes. The ideas in the derivation of Eq. (6.8) can be applied to our multidimensional rate expression, Eq. (6.10), to show that within the harmonic approximation,

$$k_{A \to B} = \frac{v_1 \times v_2 \times \cdots \times v_N}{v_1^\dagger \times \cdots \times v_{N-1}^\dagger} \exp\left(-\frac{\Delta E}{k_B T}\right). \qquad (6.15)$$

Here, ΔE is the same activation energy that was defined for the one-dimensional theory, v_i are the vibrational frequencies associated with the energy minimum, and $v_j^\dagger$ are the real vibrational frequencies associated with the transition state.

Equation (6.15) *is the main result from this section.* Even if you cannot reproduce the derivation above on a moment's notice, you should aim to remember Eq. (6.15). This result says that if you can find the energy of a minimum and a transition state and the vibrational frequencies associated with these states, then you can calculate the rate for the process associated with that transition state. Even though the activation energy in Eq. (6.15) is defined by the minimum energy path, the overall transition theory rate takes into account contributions from other higher energy trajectories.

If we use the data in Table 6.2 in Eq. (6.15) for the hopping of Ag on Cu(100), we obtain a prefactor of 1.57×10^{12} Hz. This value is quite similar to the result we obtained earlier from one-dimensional TST, which gave a prefactor of 1.94×10^{12} Hz. You can now see from Fig. 6.5 that the extra information that is gained by applying multidimensional TST is not large, particularly when we remember that we can estimate the prefactor as being 10^{12}–10^{13} Hz without doing any calculations at all.

A final point related to the vibrational modes of a transition state is whether we can use them to prove that we have correctly located a transition state. This may seem like a pedantic point, but we will see in the following sections that locating transition states in DFT calculations can be significantly more complicated than finding local energy minima, so being able to know when a calculation of this type has been successful is important. The existence of one imaginary frequency is a necessary condition for any configuration to be a transition state. This means that if you calculate the vibrational modes for a configuration and find more than one (or no) imaginary frequencies, then you can definitively conclude that you have *not* found a transition state. However, the Hessian matrix can be defined for any set of coordinates, and in general there is a region around the transition state that has one negative eigenvalue. This means that the existence of exactly one imaginary frequency cannot be used to prove that the configuration of interest is a transition state. To guarantee you have found a transition state, it is necessary to show both that there is exactly one imaginary frequency among the normal modes *and* that the configuration is a critical point on the energy surface.

6.3 FINDING TRANSITION STATES

Among all the details we have seen so far in this chapter, there are two central ideas that we hope you will remember long after putting this book down:

Idea 1: DFT calculations can be used to define the rates of chemical processes that involve energy barriers.

Idea 2: The general form for the rate of a chemical process of this kind is $k \sim \exp(-\Delta E / k_B T)$, where the activation energy, ΔE, is the energy difference between the energy minimum and the transition state associated with the process of interest.

If you accept these two ideas, then the critical steps in defining chemical rates are to find the energy minima and transition states. For the example we presented above, we could locate the transition state using the symmetry of the system. In many situations, however, this intuition-based approach is not adequate. Instead, it is important to be able to use numerical methods that will find transition states without "knowing" the answer ahead of time. In Chapter 3 we discussed how to find energy minima. Unfortunately, the iterative methods you learned there typically cannot be applied to transition states because they generate a series of configurations that move "downhill" toward a minimum. To see why these methods cannot (in general) find transition states, think about what would happen if you started an iterative minimization method from some point near a transition state on the two-dimensional surface shown in Fig. 6.2.

Because locating transition states is so important in defining chemical rates, many numerical methods have been developed for this task. There are two points that are important to consider when choosing which of these methods are the most useful. First, plane-wave DFT calculations readily provide first-derivative information from the energy surface, but not second derivatives. This means that so-called mode-following methods that find transition states by following low curvature directions from energy minima are not easily implemented in plane-wave calculations. These methods are often the methods of choice in localized basis set quantum chemistry calculations, where second-derivative information can be obtained. A second point to consider is that computing the energy of even a single configuration using DFT (or other quantum chemistry methods) typically requires considerable computational resources, so transition-state finding methods must converge to transition states in a small number of iterative steps.

The method that is most widely used for finding transition states in plane-wave DFT calculations is the *nudged elastic band* (NEB) method. This method was developed by Hannes Jónsson and co-workers as a refinement of earlier "chain-of-states" methods. The aim of a chain-of-states calculation is to define the minimum energy path (MEP) between two local minima. This idea is illustrated in Fig. 6.5, which is a contour diagram of a two-dimensional energy surface. In the figure, the MEP is represented by the smooth line connecting the reactant and product minima via the transition state. You can think of the locations labeled 0–7 in this figure as different sets of coordinates for the atoms in the system in which we are interested. We will refer to these locations as images. Calculating the energy at each image in the figure would require performing eight independent DFT calculations.

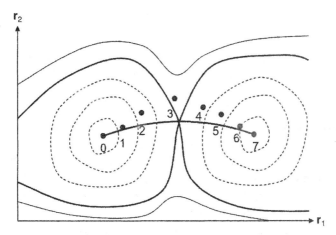

Figure 6.6 Schematic illustration of a two-dimensional energy surface with two local minima separated by a transition state. The dark curves are energy contours with energy equal to the transition-state energy. The transition state is the intersection point of the two dark curves. Dashed (solid) curves indicate contours with energies lower (higher) than the transition-state energy. The MEP is indicated with a dark line. Filled circles show the location of images used in an elastic band calculation.

If you remember that the forces on the atoms in any configuration are defined by $\mathbf{F} = -\nabla E(\mathbf{r})$, where $\mathbf{r}$ is the set of coordinates of the atoms, then the images in Fig. 6.6 can be separated into two groups. Images 0 and 7 in the figure are located at local minima, so $\mathbf{F} = 0$ for these images. For all the other images, the forces on the atoms are nonzero. A minimum energy path is defined in terms of the forces on images along the path: a path is an MEP only if the forces defined by any image along the path are oriented directly along the path. If you draw the vector defined by the forces on image 3 in Fig. 6.6, you can see that the images in this figure do not define an MEP.

6.3.1 Elastic Band Method

We will first introduce the idea of finding a minimum energy path using the elastic band method. This method is based on the concept that images along an MEP should use the lowest amount of energy to define a path between the two minima and that the images should be evenly spaced along the path. These two ideas can be expressed mathematically for a set of images $\mathbf{r}_0, \mathbf{r}_1, \ldots, \mathbf{r}_P$ by defining the objective function

$$M(\mathbf{r}_1, \mathbf{r}_2, \ldots, \mathbf{r}_P) = \sum_{i=1}^{P-1} E(\mathbf{r}_i) + \sum_{i=1}^{P} \frac{K}{2}(\mathbf{r}_i - \mathbf{r}_{i-1})^2. \qquad (6.16)$$

Here, $E(\mathbf{r}_i)$ is the total energy of the ith image, and K is a constant that defines the stiffness of the harmonic springs (the "elastic bands") connecting adjacent images. The objective function does not include the energy of image 0 or P because those images are held fixed at the energy minima. You should convince yourself that the minimization of this objective function moves all the images closer to the true MEP.

One problem with the elastic band method is illustrated in Fig. 6.7. If the penalty in the objective function for "stretching" one or more of the springs is too low, then images tend to slide downhill, away from the transition state, which of course is precisely the location we want to the highest image to approximate. This difficulty can be reduced, at least in principle, by finding an appropriate stiffness for the spring constants.

The elastic band method has a second problem known as "corner cutting," which is illustrated in Fig. 6.8. This figure shows a set of images that minimizes the objective function defined above that does not define a path going through the transition state. The difficulty here is that the true MEP follows a longer path than the path located by the elastic band method, leading to an overestimate of the activation energy for the process.

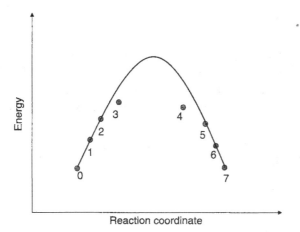

Figure 6.7 Typical outcome if the images from Fig. 6.6 are adjusted using the elastic band method with spring constants that are too small. The curve shows the energy profile for the true MEP between the two energy minima.

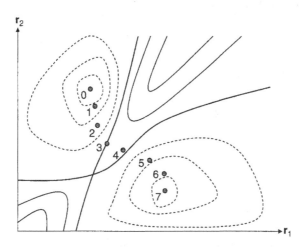

Figure 6.8 Similar to Fig. 6.6, but showing a set of images from an elastic band calculation that exhibits "corner cutting."

6.3.2 Nudged Elastic Band Method

The nudged elastic band method is designed to improve upon the elastic band method. This method is best introduced by considering how a set of images can be adjusted to move them toward the MEP. For each image, we can use DFT to compute the force acting on the system, $\mathbf{F}_i = -\nabla E(\mathbf{r}_i)$. From the current positions of the images, we can also estimate the direction of the path defined by the images. A useful estimate for this direction is to define the path direction for image i, $\hat{\tau}_i$, as a unit vector pointing along the line defined by the two adjacent images, $\mathbf{r}_{i+1} - \mathbf{r}_{i-1}$. The images will satisfy the definition of an MEP given above if the component of the force not pointing along the path direction is zero, that is, $\mathbf{F}_i^{\perp} = \mathbf{F}_i - (\mathbf{F}_i \cdot \hat{\tau}_i)\hat{\tau}_i = 0$. This description suggests a simple strategy for adjusting the images—move each of them "downhill" along the direction defined by $\mathbf{F}_i^{\perp}$. If we want to include harmonic springs between images, then we also need to include the spring forces in defining this downhill direction. A useful way to do this is to define

$$\mathbf{F}_{i,\text{update}} = \mathbf{F}_i^{\perp} + \mathbf{F}_{i,\text{spring}} = \mathbf{F}_i^{\perp} + K(|\mathbf{r}_{i+1} - \mathbf{r}_i| - |\mathbf{r}_i - \mathbf{r}_{i-1}|). \qquad (6.17)$$

This definition is appropriate for using the elastic band method.

As we pointed out above, for the forces defined by the energy surface, we are only interested in the components of these forces perpendicular to the MEP. The observation that leads to the NEB method is that for the forces defined by the elastic band springs, we are only interested in the force components that point along the direction of the MEP. That is, we want the spring forces to act only to keep the images evenly spread out along the path, and we do not want the spring forces to pull the images away from the MEP. To do this, we define

$$\mathbf{F}_{i,\text{spring}}^{\|} = (\mathbf{F}_{i,\text{spring}} \cdot \hat{\tau}_i)\hat{\tau}_i \qquad (6.18)$$

and then update the positions of each image using

$$\mathbf{F}_{i,\text{update}} = \mathbf{F}_i^{\perp} + \mathbf{F}_{i,\text{spring}}^{\|}. \qquad (6.19)$$

If all the images in the calculation lie on an MEP, then this update force is zero for every image and the calculation has converged.

The mathematical definition of the NEB method is important, but the main features of this method can be stated without these details. These include:

1. The aim of an NEB calculation is to define a series of atomic coordinates (images) that define an MEP connecting two minima on the energy surface.
2. The NEB method finds an MEP using a force projection scheme in which real forces (those resulting from the potential energy) act perpendicular to the band and spring forces act parallel to the band.

3. The NEB is an iterative minimization method, so it requires an initial estimate for the MEP. The convergence rate of an NEB calculation will depend strongly on how close the initial estimate of the path is to a true MEP.

4. Using more images to represent the path between two minima will give a more accurate description of the MEP, but this occurs at the expense of increasing the computational burden. During every iteration in an NEB calculation, a DFT calculation must be performed for every image (except the end points lying at energy minima).

There are several variations on the basic NEB method. These include more accurate ways of estimating the direction of the MEP from the current images and techniques for using springs of variable strength along the MEP to increase the density of images near the transition state. One especially useful idea is the climbing image nudged elastic band method, which adapts the basic NEB method so that when the calculation converges it places one of the images precisely at the transition state. You should become familiar with which options for these calculations are available in the software package you use.

6.3.3 Initializing NEB Calculations

There is one aspect of performing NEB calculations that is so important that we need to discuss it in some detail, namely, getting the calculation started. We will discuss this in the context of the same example we discussed in Sections 6.1 and 6.2: diffusion of an Ag atom on Cu(100). For the sake of specificity, we will consider using a NEB calculation with 8 images plus the 2 end points. In our DFT calculations, we used a slab of Cu(100) with 4 layers and 8 Cu atoms in each layer. The top 2 Cu layers were allowed to relax in every calculation, so each image had $M = 17$ unconstrained atoms. This means that a single image in an NEB calculation is defined by

$$\mathbf{r}_i = (r_{1,i}, r_{2,i}, \ldots, r_{3M,i}) \tag{6.20}$$

since each unconstrained Cu atom and the Ag atom has 3 degrees of freedom. Here, r_1, r_2, and r_3 are the spatial coordinates of a single nucleus. To begin an NEB calculation using 8 images, we must specify these $3M$ coordinates separately for each image.

We begin by specifying the two end states for the calculation, $\mathbf{r}_0$ and $\mathbf{r}_9$. These images can be found by a standard energy minimization calculation; we place the Ag atom in a fourfold site and allow it and the surface atoms to relax. The usual way to define the remaining images for the NEB calculation is by linear interpolation between the end points. That is, to generate 8 images between the fixed end points, we define the $3M$ coordinates of image i by

$$r_{j,i} = r_{j,0} + (r_{j,9} - r_{j,0})\left(\frac{i}{9}\right). \tag{6.21}$$

Linear interpolation has the great advantage of being very easy to use—almost no work is necessary to define the initial images when using this method. The simple illustrations in Figs. 6.5 and 6.7 were chosen in part to show that linear interpolation of the images is a good approximation in some examples but not in others. In Fig. 6.5, linear interpolation would give a reasonable initial estimate for the MEP. In Fig. 6.7, in contrast, the true MEP is strongly curved, so linear interpolation would give a less accurate estimate for the MEP.

It may be tempting to conclude that linear interpolation is almost exact for our example of Ag on Cu(100) since the MEP really is a straight line between two fourfold sites if viewed from the top of the surface as in Fig. 6.1. This is not correct because it ignores the other degrees of freedom that are needed to define the MEP. In particular for this example, the height of the Ag atom above the Cu surface is the same in the two minima. This means that in the images defined by Eq. (6.21), the Ag atom lies at the same height above the Cu surface in every image. Since, in fact, the height of the Ag atom over the Cu surface does change along the MEP, linear interpolation does not give the exact MEP, even for this symmetric example.

The results of using an NEB calculation to describe Ag hopping on Cu(100) are shown in Fig. 6.9. In this calculation, 27 iterations of the NEB method were performed before reaching a converged result. As with any iterative

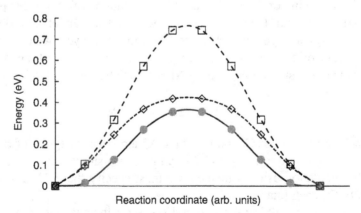

Figure 6.9   Energy profiles computed using NEB for Ag diffusion on Cu(100). Open squares show the energies associated with the initial images defined via linear interpolation. Open diamonds show the initial energies defined via linear interpolation but with the height of the Ag atom adjusted as described in the text. The converged final state for both calculations is shown with filled circles.

optimization calculation, the number of iterations needed to achieve convergence can be reduced if the quality of the initial estimate of the solution is improved. For constructing initial estimates for the images along an MEP, this often means that it is worthwhile to adjust the estimates with some physical or chemical intuition about the process being described. In the present example, we might anticipate that the Ag atom will be slightly higher on the surface as it goes over the bridge site than it is in the fourfold sites. To construct initial images that include this simple idea, we took the configurations defined by linear interpolation and increased the height of the Ag atom by 0.1 Å for images 2 and 7, 0.15 Å for images 3 and 6, and 0.2 Å for images 4 and 5. The energies associated with these initial states are shown with open diamonds in Fig. 6.9. Repeating the NEB calculation beginning from this initial estimate yielded a converged result in 24 iterations. Although this is a relatively modest improvement in computational effort, it illustrates the idea that putting a little effort into improving upon simple linear interpolation can improve the convergence rate of NEB calculations.

Comparing Fig. 6.9, which shows the MEP determined from NEB calculations, with Fig. 6.3, which shows the MEP determined from symmetry considerations, points to an important shortcoming of our NEB results: none of the images in the NEB calculation lie precisely at the transition state. Image 4 and image 5 both lie 0.35 eV above the end points of the calculated MEP, a result that slightly underestimates the true activation energy. For the particular example we have been using, this problem could easily have been avoided by using an odd number of images in the NEB calculation since we know from symmetry that the transition state must lie precisely in the middle of the reaction coordinate. We chose not to do this, however, to highlight the important point that in almost all situations of practical interest, the location of the transition state *cannot* be determined by symmetry alone.

Because finding a precise value for the activation energy is so important in using transition state theory (see the discussion of Fig. 6.5 above), NEB calculations performed as we have described them need to be followed by additional work to find the actual transition state. There are at least two ways to tackle this task. First, variations of the NEB method have been developed in which the converged result not only spaces images along the MEP but also has one of the images at the transition state. Information about this method, the climbing image NEB method, is given in the further reading at the end of the chapter. Another simple approach relies on the fact that some geometry optimization methods will converge to a critical point on an energy surface that is a transition state or saddle point if the initial estimate is sufficiently close to the critical point. Said more plainly, geometry optimization can be used to optimize the geometry of a transition state if a good approximation for the geometry of the state can be given.

The images on the MEP calculated by the NEB calculations above provide a straightforward way to approximate the geometry of the transition state. Fitting a smooth curve through the NEB-calculated images such as the data shown in Fig. 6.9 gives a well-defined maximum in energy. In our simple example, this maximum occurs exactly half way between images 4 and 5 because of the symmetry of the problem. If we create a configuration by linear interpolation between the two images nearest the maximum in the fitted curve at the value of the reaction coordinate defined by the maximum, this will typically give a good estimate for the transition-state geometry. Performing a geometry optimization starting from this configuration converges quickly to a transition state with a corresponding activation energy of 0.36 eV. We reiterate that this process of refining the configuration associated with the transition state is much more important in examples where symmetry cannot be used to locate the transition state.

6.4   FINDING THE RIGHT TRANSITION STATES

Using the techniques described in this chapter, you may identify the geometry of a transition state located along the minimum energy path between two states and calculate the rate for that process using harmonic transition state theory. However, there is a point to consider that has not been touched on yet, and that is: how do you know that the transition state you have located is the "right" one? It might be helpful to illustrate this question with an example.

A process that is very important for phenomena such as crystal growth and epitaxy is the diffusion of metal atoms across a metal surface. When the atom that is diffusing across the surface is of the same type as the atoms that make up the crystal, this process is known as self-diffusion. The conventional view of self-diffusion is of an atom that hops from local minimum to local minimum on top of the surface; in this picture the diffusing atom is called an adatom. This process, illustrated in Fig. 6.10, is the reaction path we assumed for an Ag atom moving about on a Cu(100) surface. We can also adopt this hopping reaction path to describe diffusion of a Cu atom on the Cu(100) surface. Performing NEB calculations for Cu hopping on Cu(100) in the same way as the Ag/Cu(100) calculations described above predicts that the activation energy for this process is 0.57 eV.

The simple hopping mechanism, however, is not the only possible way for atoms to move on Cu(100). Figure 6.11 illustrates another possible diffusion process known as an exchange mechanism. In this mechanism, the adatom (shown in dark grey), replaces an atom that is initially a surface atom (shown in light grey) while the surface atom pops up onto the surface,

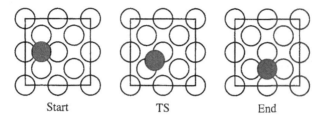

Start   TS   End

Figure 6.10 Illustration of a Cu adatom hopping from a hollow site to an adjacent hollow site on Cu(100). The diagram on the left shows the adatom (grey) before the hop, the middle shows the transition state (TS), and the diagram on the right shows the adatom after the hop.

becoming an adatom. This process has the net effect of moving an adatom between two sites on the surface. It was first detected on the (110) surfaces of certain fcc metals, but has also been observed on the (100) surface of metals including Al and Pt.

If we want to understand the diffusion of Cu adatoms on Cu(100), we clearly should compare the rates of the two diffusion mechanisms shown in Figs. 6.10 and 6.11. What do the NEB calculations described in the previous section tell us about the possible occurrence of exchange hopping? The answer is simple: nothing! Because we start our NEB calculations by supplying an initial path that interpolates between the start and end states in Fig. 6.10, our result cannot even give a hint that paths like the one in Fig. 6.11 exist, much less give any detailed information about these paths. In more general terms: *NEB calculations are local not global optimization calculations, so an NEB calculation can give accurate information about a particular transition state, but it cannot indicate whether other transition states related to the same initial state also exist.*

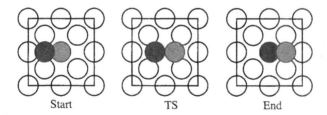

Start   TS   End

Figure 6.11 Illustration of the exchange mechanism for self-diffusion on the Cu(100) surface. The diagram on the left shows the Cu adatom (dark grey) in a hollow site and highlights one of the surface atoms in light grey. In the transition state, shown in the middle, both Cu atoms lie above the plane of the surface. The diagram on the right shows the result after exchange, where the original adatom has become part of the surface and the original surface atom has become an adatom.

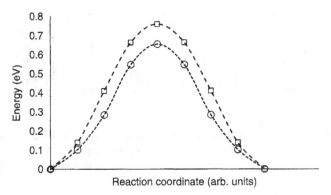

Figure 6.12   Comparison of NEB results for exchange (squares) and direct adatom hopping (circles) mechanisms for self-diffusion of Cu on Cu(100).

To explore whether multiple transition states exist for a particular process, NEB calculations must be performed independently for a variety of possible activated processes. It is reasonably straightforward to use NEB to examine exchange hopping of Cu on Cu(100). We performed calculations using seven images (plus the end points) created using linear interpolation from the start and end states shown in Fig. 6.11. In this case the middle image converges to the transition state by symmetry. Similar calculations were performed for the hopping mechanism shown in Fig. 6.10. The results are shown in Fig. 6.12. The main observation from these calculations is that the activation energy for the exchange mechanism, 0.67 eV, is higher than the barrier for direct hopping, 0.57 eV. If we assume that the Arrhenius prefactors for these two processes are similar, these barriers indicate that the direct hopping is about 50 times faster than exchange hopping at 300 K. Even though an exchange hop moves the adatom further on the surface than direct hopping (cf. Figs. 6.10 and 6.11), this result indicates that the net contribution of exchange hopping to Cu adatom diffusion on Cu(100) is small.

Do we now have a complete description of Cu adatom diffusion on Cu(100)? No, there may be other activated processes that we have not yet considered that have rates comparable to the rate of direct hopping. For example, exchange diffusion could occur by involving multiple surface atoms leading to the net movement of an adatom to a site further from the initial adatom than the "simple" exchange mechanism shown in Fig. 6.11. It is important in any situation like this to consider a range of possible events, a process that must be guided by physical and chemical intuition. Even after a number of possible processes have been analyzed based on NEB calculations, it is only rigorously possible to use the lowest activation energy found from these calculations as a upper bound on the true activation energy for a process starting from an initial state of interest.

6.5 CONNECTING INDIVIDUAL RATES TO OVERALL DYNAMICS

You have now learned about how to use DFT calculations to compute the rates of individual activated processes. This information is extremely useful, but it is still not enough to fully describe many interesting physical problems. In many situations, a system will evolve over time via many individual hops between local minima. For example, creation of catalytic clusters of metal atoms on metal oxide surfaces involves the hopping of multiple individual metal atoms on a surface. These clusters often nucleate at defects on the oxide surface, a process that is the net outcome from both hopping of atoms on the defect-free areas of the surface and in the neighborhood of defects. A characteristic of this problem is that it is the long time behavior of atoms as they move on a complicated energy surface defined by many different local minima that is of interest.

As a simplified example, let us extend our example of Ag hopping on Cu(100) to think about a Cu(100) surface that is doped at random positions with a small concentration of Pd atoms. To draw an analogy with the situation outlined above, you can think of the Pd atoms as defining "defects" on the surface. This hypothetical surface is shown in Fig. 6.13. If we classify the fourfold hollow sites on the surface using only the four surface atoms surrounding each site, then there are just two kinds of surface sites.[‡] By the same logic, the local hops of an Ag atom on the surface can be described using the four different rates shown in Fig. 6.13, $k_1, \ldots, k_4$. Using the DFT methods you have learned so far in this chapter, you could determine each of these rates reasonably easily.

There are a host of physical questions that cannot be easily answered just by knowing the rates listed in Fig. 6.13. For example, once an Ag atom is deposited on the surface, how long will it be (on average) before that Ag atom visits a site adjacent to a Pd surface atom? How many different Pd surface sites will an Ag atom visit per unit time on the surface? What is the net diffusion coefficient of Ag atoms on this surface? To answer these questions, we need a tool to describe the evolution in time of a set of Ag atoms on the surface.

The dynamics of systems such as this one can be followed using a method called *kinetic Monte Carlo* (kMC) The idea behind this method is straightforward: If we know the rates for all processes that can occur given the current configuration of our atoms, we can choose an event in a random way that is consistent with these rates. By repeating this process, the system's time evolution can be simulated.

[‡]This approximate approach cannot capture everything that can happen. For example, the presence of Pd atoms as next nearest neighbors to a surface site will in general change the energy of an adatom at that site.

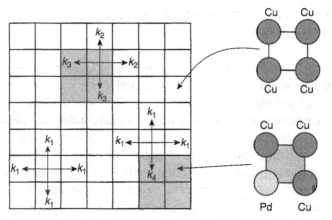

Figure 6.13 Schematic illustration of a lattice model for diffusion of Ag atoms on Pd-doped Cu(100). The diagrams on the right show the fourfold surface sites in terms of the four surface atoms defining the site. The section of the surface shown on the left includes two well-separated Pd atoms in the surface. The Pd atoms are located at the centers of the two grey squares in the diagram on the left.

It is easiest to describe the kMC method by giving a specific example of an algorithm for the system we have defined above. Let us imagine we have a large area of the surface divided into fourfold sites of the two kinds shown in Fig. 6.13 and that we have computed all the relevant rates. We will denote the largest of all the rates in our catalog of rates as k_{max}. To begin, we deposit N Ag atoms at random positions on the surface. Our kMC algorithm is then the following:

1. Choose 1 Ag atom at random.
2. Choose a hopping direction (up, down, left, or right) at random. Using our catalog of rates, look up the rate associated with this hop, k_{hop}.
3. Choose a random number, ε, between 0 and 1.
4. If $\varepsilon < k_{hop}/k_{max}$, move the selected Ag atom in the selected direction. Otherwise, do not move the atom.
5. Regardless of the outcome of the previous step, increment time by $\Delta t = 1/(4Nk_{max})$.
6. Return to step 1.

We can verify that this algorithm will generate a physically correct sequence of events by thinking about two simple situations. First, if the surface is pure Cu, then every local hopping rate has the same rate, k, so the hops in step 4 of the algorithm would always be performed. The average time between hops for any individual atom in this case will be $\Delta t = \frac{1}{4} k$. This means that the atoms hop on the surface at the correct rate. Second, if an atom can hop out of a site along two directions with different local rates, k_{high} and k_{low}, then the ratio of hops along the fast direction to hops along the slow direction is k_{high}/k_{low}. The same argument can be applied to two atoms sitting in different sites. These ideas show that different processes occur with the correct ratio of rates and that the absolute rate of the processes is correct—in other words this algorithm defines physically correct dynamics. A kMC simulation is intrinsically stochastic, so two simulations cannot be expected to generate the same sequence of events, even if they begin from the same initial conditions. It is typically necessary to average over the evolution of multiple simulations to get reliable average information.

There are two important drawbacks of kMC simulations that should be kept in mind. The first is related to the numerical efficiency of our algorithm. If the catalog of rates includes a wide range of absolute rates, then the ratio k_{hop}/k_{max} in step 4 of the algorithm can be very small for the slowest rates. For example, if the activation energies of the slowest and fastest rates are 1.0 and 0.2 eV and the rates for these processes have the same prefactors, then this ratio is $\sim 10^{-13}$ at room temperature and $\sim 10^{-7}$ at 300°C. This means our algorithm is highly inefficient; many millions of iterations of the algorithm may pass with no change in the state of our system. This problem is relatively simple to solve by using one of a variety of more sophisticated kMC algorithms that have been developed over the years. The further reading section at the end of the chapter can point you to literature about these methods.

The second drawback of kMC is more serious: kMC results can only be correct if the catalog of rates used to define the method is complete. To be specific, let us return to the diffusing Ag atoms in the example above. Our algorithm is well-defined if the Ag atoms on the surface are isolated, but what happens if two Ag atoms are found in two neighboring surface sites? We need to add information to our lattice model to define what can happen in this situation. From a physical point of view, we do not want an Ag atom to hop into a site that is already filled by another Ag atom. This is easy enough to achieve by simply defining the hopping rate to be zero if the site the atom is hopping into is already filled. But how fast will one of the Ag atoms hop away from an adjacent pair of atoms? It is very likely that these rates are different (and lower) than the rates for an isolated Ag atom on the surface since it is typically energetically favorable for metal adatoms to group together on a surface. We could include these rates in our kMC algorithm if we take the time to use DFT calculations to compute the rates. You can see from Fig. 6.13 that two adjacent Ag atoms can sit on a variety of sites. To be complete we would need to compute the rates for each of these situations separately. Similarly, if we are considering random doping of the surface by Pd atoms, then there will be locations on the surface where two Pd atoms are adjacent in the surface layer. To account for this (and larger clusters of Pd atoms in the surface), the rates of all surface processes on each possible arrangement of surface atoms should be considered.

Even if all of the processes just listed were treated with DFT, our model would still be incomplete. As time passes in our kMC simulations, clusters of Ag may form on the surface. By the assumptions of our model, these clusters have every Ag atom sitting in a fourfold site on the underlying surface. But nature is not restricted by this assumption. It is not uncommon for small metal clusters on surfaces to be three dimensional in nature. Treating this situation in a detailed way with our kMC scheme would be very difficult—we would have to consider all the possible processes that could allow Ag atoms to climb up onto or down from existing Ag clusters and compute their rates, then modify our algorithm to allow for three-dimensional movement. These complications do not stop kMC simulations from being useful, but they do mean that the results of these simulations should always be interpreted by understanding what phenomena can and cannot be probed because of the structure of the underlying model.

6.6   QUANTUM EFFECTS AND OTHER COMPLICATIONS

Throughout this chapter we have focused on chemical processes that involve an activation barrier. As long as the height of this barrier is significant relative to k_BT, transition-state theory can supply a useful description of the rate of the overall process. But there are situations where transition state theory is not adequate, and it is helpful to briefly consider two common examples.

6.6.1   High Temperatures/Low Barriers

The first situation where transition state theory can be inaccurate is at temperatures where the activation barriers are similar or even small relative to k_BT. As an example, think about the example above of an Ag atom sitting on a Cu(100) surface. One way to get the Ag atom onto the surface would be to evaporate Ag atoms in a vacuum chamber containing a Cu(100) crystal. The process of an Ag atom arriving from the gas phase and sticking to the crystal is a *barrierless process*: it is much more energetically favorable for the Ag atom to be on the surface than in the gas phase, and it can be moved from the gas phase to the surface following a path that monotonically decreases in energy. The formulation of transition state theory we have been exploring simply cannot describe this situation. Theories that can be applied to situations such as this have been extensively developed in the study of chemical kinetics—the Further Reading section at the end of the chapter indicates several places to explore these ideas.

6.6.2 Quantum Tunneling

A rather different failure of transition-state theory can occur at very low temperatures. Transition-state theory is built around the idea that transitions between energy minima take place by the system gaining energy and going over a barrier. Quantum mechanics also allows another type of transition in which particles pass through barriers even though their energy is considerably less than the energy needed to go over the barrier; this is the phenomenon of quantum tunneling. Tunneling rates decrease rapidly with increasing mass of the moving particle, so tunneling is most important for transitions involving hydrogen.

Fortunately, it is relatively simple to estimate from harmonic transition-state theory whether quantum tunneling is important or not. Applying multidimensional transition-state theory, Eq. (6.15), requires finding the vibrational frequencies of the system of interest at energy minimum A $(v_1, v_2, \ldots, v_N)$ and transition state $(v_1^\dagger, \ldots, v_{N-1}^\dagger)$. Using these frequencies, we can define the zero-point energy corrected activation energy:

$$\Delta E_{zp} = \left(E^\dagger + \sum_{j=1}^{N-1} \frac{h v_j^\dagger}{2} \right) - \left(E_A + \sum_{i=1}^{N} \frac{h v_i}{2} \right). \tag{6.22}$$

We also know that the one imaginary frequency associated with the transition state has the form $i v_{\mathrm{Im}}^\dagger$. These quantities can be combined to form a crossover temperature, T_c, defined by

$$T_c = \frac{h v_{\mathrm{Im}}^\dagger E_{zp} / k_B}{2\pi E_{zp} - h v_{\mathrm{Im}}^\dagger \ln 2}. \tag{6.23}$$

Below this crossover temperature, tunneling can make a significant contribution to the total rate, while above this temperature tunneling can be neglected. Fermann and Auerbach have developed a correction to harmonic transition-state theory that can be used to estimate tunneling rates below T_c using information that is already available when applying the classical theory.[1]

6.6.3 Zero-Point Energies

A final complication with the version of transition-state theory we have used is that it is based on a classical description of the system's energy. But as we discussed in Section 5.4, the minimum energy of a configuration of atoms should more correctly be defined using the classical minimum energy plus a zero-point energy correction. It is not too difficult to incorporate this idea into transition-state theory. The net result is that Eq. (6.15) should be modified to include zero-point energies to read

$$k_{A \to B} = \frac{\prod\limits_{i=1}^{N} f(h\nu_i/2k_B T)}{\prod\limits_{j=1}^{N-1} f(h\nu_j^\dagger/2k_B T)} \frac{\nu_1 \times \nu_2 \times \cdots \times \nu_N}{\nu_1^\dagger \times \cdots \times \nu_{N-1}^\dagger} \exp\left(-\frac{\Delta E}{k_B T}\right), \qquad (6.24)$$

where $f(x) = \sinh(x)/x$. This expression is not as difficult to understand as it looks. At high temperatures, $f(h\nu/2k_B T) \to 1$, so Eq. (6.24) reduces exactly to the classical expression we used earlier. At low temperatures, the limiting result is even simpler:

$$k_{A \to B} = \frac{k_B T}{h} \exp\left(-\frac{\Delta E_{zp}}{k_B T}\right). \qquad (6.25)$$

Here, ΔE_{zp} is the zero-point energy corrected activation energy defined in Eq. (6.22). At intermediate temperatures, Eq. (6.24) smoothly connects these two limiting cases. In many examples that do not involve H atoms, the difference between the classical and zero-point corrected results is small enough to be unimportant.

An interesting outcome from Eq. (6.25) is that the prefactor in this Arrhenius expression is independent of the system's vibrational frequencies. In Section 6.1 we noted that this prefactor could be estimated in terms of "typical" vibrational frequencies as being 10^{12}–10^{13} s^{-1}. Equation (6.25) supports this estimate: As T varies from 100–500 K, $k_B T/h$ changes from 2×10^{12} to 1.2×10^{13} s^{-1}.

EXERCISES

1. Determine the activation energy for the diffusion of an Ag atom between adjacent threefold sites on Cu(111) using the NEB method. Note that the energy of the end points in your calculation will not be exactly equal because fcc and hcp sites on the Cu(111) surface are not identical. Compute the frequencies for this hopping process.

2. Determine the activation energy for migration of a charge-neutral vacancy in bulk Si by removing one Si atom from a 64-atom supercell of the bulk material. Before performing any NEB calculations, think carefully about whether it matters whether any atoms moves across the boundary of your supercell between the initial and final state. Using your calculated activation energy and a standard estimate for a transition-state prefactor, predict the hopping rate of charge-neutral vacancies in Si.

3. Hydron atoms readily dissolve into bulk Pd, where they can reside in either the sixfold octahedral or fourfold tetrahedral interstitial sites. Determine the classical and zero-point corrected activation energies for H hopping between octahedral and tetrahedral sites in bulk Pd. In calculating the activation energy, you should allow all atoms in the supercell to relax; but, to estimate vibrational frequencies, you can constrain all the metal atoms. Estimate the temperature below which tunneling contributions become important in the hopping of H atoms between these two interstitial sites.

4. Calculate the activation energy for diffusion of a Pt adatom on Pt(100) via direct hopping between fourfold sites on the surface and, separately, via concerted substitution with a Pt atom in the top surface layer. Before beginning any calculations, consider how large the surface slab model needs to be in order to describe these two processes. Which process would you expect to dominate Pt adatom diffusion at room temperature?

5. As an example of why linear interpolation is not always a useful way to initialize an NEB calculation, consider the molecule HCN in the gas phase. This molecule can rearrange to form CNH. Optimize the structures of HCN and CNH, then use these states to examine the bond lengths in the structures that are defined by linear interpolation between these two structures. Why are the intermediate structures defined by this procedure not chemically reasonable? Construct a series of initial images that are chemically reasonable and use them in an NEB calculation to estimate the activation energy for this molecular isomerization reaction.

REFERENCE

1. J. T. Fermann and S. Auerbach, Modeling proton mobility in acidic zeolite clusters: II. Room temperature tunneling effects from semiclassical theory, *J. Chem. Phys.* **112** (2000), 6787.

FURTHER READING

The field of chemical kinetics is far reaching and well developed. If the full energy surface for the atoms participating in a chemical reaction is known (or can be calculated), sophisticated rate theories are available to provide accurate rate information in regimes where simple transition state theory is not accurate. A classic text for this field is K. J. Laidler, *Chemical Kinetics*, 3rd ed., Prentice Hall, New York, 1987. A more recent book related to this topic is I. Chorkendorff and J. W. Niemantsverdriet, *Concepts of Modern Catalysis and Kinetics*, 2nd ed., Wiley-VCH, Weinheim, 2007. Many other books in this area are also available.

To explore the connection between statistical mechanics and TST, see D. Chandler, *Introduction to Modern Statistical Mechanics*, Oxford University Press, Oxford, UK, 1987, and D. Frenkel and B. Smit, *Understanding Molecular Simulation: From Algorithms to Applications*, 2nd ed., Academic, San Diego, 2002.

For a description of the NEB method and a comparison of the NEB method with other chain of states methods for determining transition states without the use of the Hessian matrix, see D. Sheppard, R. Terrell, and G. Henkelman, *J. Chem. Phys.* **128** (2008), 134106. The climbing image NEB method is described in G. Henkelman, B. P. Uberuaga, and H. Jónsson, *J. Chem. Phys.* **113** (2000), 9901–9904. Improvements to the NEB method including a better estimation of the tangent to the MEP are described in G. Henkelman and H. Jónsson, *J. Chem. Phys.* **113** (2000), 9978–9985.

For detailed examples of exchange mechanisms in atomic diffusion on surfaces, see M. L. Anderson, M. J. D'Amato, P. J. Feibelman, and B. S. Swartzentruber, *Phys. Rev. Lett.* **90** (2003), 126102, and P. J. Feibelman and R. Stumpf, *Phys. Rev. B* **59** (1999), 5892–5897. An example of multiple mechanisms that must be considered to describe atomic diffusion and dopant clustering in bulk materials is given in B. P. Uberuaga, G. Henkelman, H. Jonsson, S. T. Dunham, W. Windl, and R. Stumpf, *Phys. Stat. Sol. B* **233** (2002), 24–30.

The kinetic Monte Carlo (kMC) algorithm we have described is straightforward to derive and implement but is not necessarily the most efficient of the various algorithms that exist. For a careful review of the properties and efficiency of different kMC algorithms, see J. S. Reese, S. Raimondeau, and D. G. Vlachos, *J. Comput. Phys.* **173** (2001), 302–321.

A review of using DFT methods in combination with kinetic Monte Carlo for modeling active sites on metal catalysts, see M. Neurock, *J. Catalysis* **216** (2003), 73–88.

APPENDIX   CALCULATION DETAILS

The calculations in this chapter are based on two systems: an Ag adatom diffusing from the hollow to the bridge site on the Cu(100) surface and Cu self-diffusion on Cu(100). For all the calculations, an asymmetric four-layer Cu slab model was used with the bottom two layers fixed. The dimensions of the supercell in the plane of the surface were fixed to the DFT-optimized Cu lattice constant. The supercell was a c(4×4) surface unit cell containing eight atoms per layer. The vacuum space was greater than 28 Å. The PBE GGA functional was employed with a cutoff energy of 380 eV for all calculations. For energy minimizations as well as for the nudged elastic band calculations, a conjugate-gradient algorithm was used to relax all unconstrained ions. The Methfessel–Paxton smearing method was applied with a width of

0.1 eV and a dipole correction was applied in the direction normal to the surface. Further details regarding specific calculations from each section of the chapter are given below.

Section 6.1   Minimizations of an Ag adatom on hollow and bridge sites of Cu(100) were performed with the parameters described above and $4 \times 4 \times 1$ k points.

Section 6.2   Frequency calculations were performed on the Ag adatom on the Cu(100) hollow and bridge sites using $4 \times 4 \times 1$ k points. The frequencies were determined via finite-difference calculations using a displacement of 0.04 Å.

Section 6.3   The nudged elastic band calculations for the Ag adatom to hop from hollow to hollow site via a minimum energy path passing through the bridge site were performed using eight images plus the two end points using $4 \times 4 \times 1$ k points for each image.

Section 6.4   NEB calculations for Cu surface diffusion were performed with seven images using $2 \times 2 \times 1$ k points.

7

EQUILIBRIUM PHASE DIAGRAMS FROM *AB INITIO* THERMODYNAMICS

The task that consumed most of our attention in Chapter 2 was using DFT calculations to predict the crystal structure of bulk copper. Once you learned the details of defining various crystal structures and minimizing their energies with DFT, this task was relatively straightforward. Specifically, we minimized the energy of bulk Cu in several candidate crystal structures (cubic, fcc, and hcp) and then compared the energy of each structure in units of energy/ atom. This calculation made a prediction in agreement with experimental observation that copper has an fcc crystal structure.

Said more rigorously, our earlier calculations make a prediction for the crystal structure of Cu at zero pressure. It is certainly important to consider how we can extend our results to more practically relevant conditions! In Section 2.5, we described how the relative stability of several possible structures of a material can be compared at nonzero pressures by calculating the cohesive energy of the material as a function of volume. This concept is very important for applications such as geophysics that deal with extremely high pressures. The vignette in Section 1.2 about planetary formation is one example of this situation.

Now imagine that sitting next to you as you read this is a small lump of Cu. Are our DFT calculations relevant for predicting the structure of this material? One crucial difference between our DFT calculations in Chapter 2 and your lump of Cu is that the latter is surrounded by billions of gas molecules

Density Functional Theory: A Practical Introduction. By David S. Sholl and Janice A. Steckel
Copyright © 2009 John Wiley & Sons, Inc.

(mainly N_2 and O_2) that can potentially react with the metal. We will simplify our discussion by thinking of N_2 as unreactive, so the Cu is surrounded by O_2 at a pressure of roughly 0.2 atm at room temperature. It is certainly possible to oxidize copper; a common mineral form of oxidized copper is Cu_2O. This means that if we want to predict the properties of Cu under ambient conditions, we need to be able to decide which of Cu and Cu_2O is the most stable in equilibrium with gas phase O_2.

It may be tempting to think that the relative stability of Cu and Cu_2O can be predicted in the same way we solved this problem when we looked at crystal structures for pure Cu. In this approach, we would minimize the energy of each material and decide which one has the lowest energy. There are two fundamental flaws with this idea. First, how should we normalize the energy of the different materials? When we were thinking of bulk Cu, using energy/atom was perfectly sensible. But this normalization cannot be correct when comparing several materials with different chemical compositions. Second, this approach does not include any information about the temperature and pressure of O_2 that we are aiming to describe. Presumably if the O_2 pressure was decreased from 0.2 atm to 2×10^{-6} atm the equilibrium state of Cu could change; so we need to use a method that can describe this process.

If we were only interested in bulk copper and its oxides, we would not need to resort to DFT calculations. The relative stabilities of bulk metals and their oxides are extremely important in many applications of metallurgy, so it is not surprising that this information has been extensively characterized and tabulated. This information (and similar information for metal sulfides) is tabulated in so-called Ellingham diagrams, which are available from many sources. We have chosen these materials as an initial example because it is likely that you already have some physical intuition about the situation. The main point of this chapter is that DFT calculations can be used to describe the kinds of phase stability that are relevant to the physical questions posed above. In Section 7.1 we will discuss how to do this for bulk oxides. In Section 7.2 we will examine some examples where DFT can give phase stability information that is also technologically relevant but that is much more difficult to establish experimentally.

7.1 STABILITY OF BULK METAL OXIDES

As an initial example of using DFT calculations to describe phase stability, we will continue our discussion of bulk metal oxides. In this chapter, we are interested in describing the thermodynamic stability of a metal, M, in equilibrium with gas-phase O_2 at a specified pressure, P_{O_2}, and temperature, T. We will assume that we know a series of candidate crystal structures for the metal and its oxide. To be specific, we will examine Cu and Cu_2O, each in the crystal structure that is known from experiments. It turns out that Ag and Ag_2O have the same structures as the corresponding copper materials, so we will look at the oxides of silver and copper at the same time.

Thermodynamically, we would like to know which material minimizes the free energy of a system containing gaseous O_2 and a solid at the specified conditions. A useful way to do this is to define the *grand potential* associated with each crystal structure. The grand potential for a metal oxide containing N_M metal atoms and N_O oxygen atoms is defined by

$$\Omega(T, \mu_O, \mu_M) = E(N_M, N_O) - TS - \mu_O N_O - \mu_M N_M, \qquad (7.1)$$

where E is the internal energy of the metal oxide, S is the material's entropy, and μ_a is the chemical potential of atomic species a. If we have a series of different materials ($i = 1, 2, 3, \ldots$), then we can use this expression to define the grand potential of each material, Ω_i.

We can interpret the internal energy in the grand potential as simply the total energy from a DFT calculation for the material. It is then sensible to compare the grand potentials of the different materials by normalizing the DFT energies so that every DFT calculation describes a material with the same total number of metal atoms. If we do this, Eq. (7.1) can be rewritten as

$$\Omega_i(T, \mu_O) = E_i - TS_i - \mu_O N_{O,i} - \Omega^M, \qquad (7.2)$$

where Ω^M is an additive constant that is the same for every material. The chemical potential of oxygen is directly defined by the temperature and pressure of gaseous O_2. If we treat O_2 as an ideal gas, then

$$\mu_{O_2} = \mu_{O_2}^o(T, p^o) + kT \ln(p_{O_2}/p_{O_2}^o), \qquad (7.3)$$

where the superscript o denotes a reference pressure, usually taken to be 1 atm. The chemical potential of molecular oxygen and atomic oxygen are related by

$$\mu_O = \tfrac{1}{2}\mu_{O_2}. \qquad (7.4)$$

The grand potential defined in Eq. (7.2) has one crucial and simple property: The material that gives the lowest grand potential is also the material that minimizes the free energy of the combination of gaseous O_2 and a solid. In other words, once we can calculate the grand potential as shown in Eq. (7.2), the thermodynamically stable state is simply the state with the lowest grand potential.

When comparing crystalline solids, the differences in internal energy between different structures are typically much larger than the entropic differences between the structures. This observation suggests that we can treat the entropic contributions in Eq. (7.2) as being approximately constant for all the crystal structures we consider.[*] Making this assumption, the grand potential we aim to calculate is

$$\Omega_i(T, \mu_O) = E_i - \mu_O N_{O,i} - \Omega_S^M. \tag{7.5}$$

Here Ω_S^M is an additive constant that is the same for every material. Because we are only interested in which material has the lowest grand potential, we can set this constant to zero without any loss of generality: $\Omega_S^M = 0$. If you now compare Eqs. (7.3), (7.4), and (7.5), you can see that we have expressed the task of finding the most stable phase of the metal oxide in terms that only involve the DFT-calculated energies of the metal and metal oxides of interest and the pressure and temperature of gaseous O_2.

The application of Eq. (7.5) for Cu and Ag is shown in Fig. 7.1. This schematic diagram shows the relative stability of M and M_2O. The numerical details are different for M = Cu and M = Ag, but the basic content of the diagram is the same in both cases. The dotted lines in Figure 7.1*a* show the grand potentials defined by Eq. (7.5). The line for the pure metal, M, has a slope of zero because $N_O = 0$ in this case. The line for the metal oxide has a negative slope. The thick curve in Fig. 7.1*a* denotes the structure with the lowest grand potential as the oxygen chemical potential is varied. At sufficiently low values of the chemical potential, the bulk metal is thermodynamically favored. For sufficiently high chemical potentials, M_2O is the favored material. It is easier to think about this result in physical terms if the oxygen chemical potential is described in terms of pressure and temperature. This gives the schematic phase diagram shown in Fig. 7.1*b*. This phase diagram agrees with the well-known trends for oxidation and reduction of metals, namely that metal oxides can be reduced to metals by holding the material at a high enough temperature or in a low enough O_2 pressure.

One feature of Fig. 7.1 is that both of the materials shown on the diagram are thermodynamically favored for some combination of temperature and O_2 pressure. This does not have to be the case. Figure 7.2 shows an extension of this grand potential diagram highlighting the thermodynamically favored materials already identified and adding information for a third oxide with

---

[*]If greater precision is required in describing entropic contributions, the vibrational density of states for each material can be calculated as briefly described in Section 5.5. The entropy due to lattice vibrations can then be calculated. For more details on this concept, see the references associated with Section 5.5.

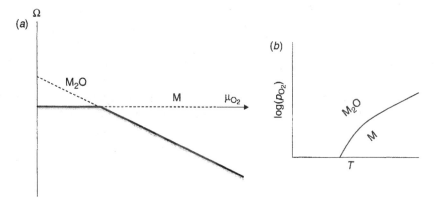

Figure 7.1 (*a*) Schematic grand potential diagram for oxidation of Cu or Ag and (*b*) the phase diagram defined by the grand potential diagram. The curves in this diagram separate regions where either the metal or the oxide is the thermodynamically favored material.

stoichiometry M_3O_2. In this example, the grand potential of the newly added oxide always lies above the grand potential of at least one of the other materials, meaning that M_3O_2 is never thermodynamically stable. This means that the phase diagram shown in Fig. 7.1*b* is still correct even when M_3O_2 is considered in addition to M and M_2O.

One important caveat that must be kept in mind when using these methods is that diagrams such as Figs. 7.1 and 7.2 can only tell us about the relative stability of materials *within the set of materials for which we performed*

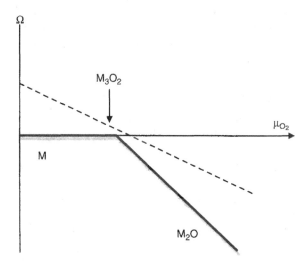

Figure 7.2 Extension of the grand potential diagram from Fig. 7.1 with the addition of a metal oxide that is not thermodynamically stable under any conditions.

DFT calculations. This situation is very similar to the one we came across in Chapter 2 where we were interested in "predicting" crystal structures. If our DFT calculations do not include one or more crystal structures that are relevant to a real material, then the phase diagram from the methods illustrated above cannot be complete.

To interpret the phase diagram in Fig. 7.1 quantitatively, we must return to Eq. (7.3) and more fully define the chemical potential. For ideal gases, the chemical potential can be rigorously derived from statistical mechanics. A useful definition of the ideal-gas chemical potential for O_2 is

$$\mu_{O_2} = E_{O_2}^{\text{total}} + \tilde{\mu}_{O_2}(T, p^o) + kT \ln(p/p^o) \tag{7.6}$$

Here, $E_{O_2}^{\text{total}}$ is the total energy of an isolated O_2 molecule at $T = 0$ K. If we neglect zero-point energies, we can obtain this energy from a simple DFT calculation of the isolated molecule.[†] The second term on the right in Eq. (7.6) is the difference in the chemical potential of O_2 between $T = 0$ K and the temperature of interest at the reference pressure. This chemical potential difference can be evaluated for O_2 and many other gases using data tabulated in the *NIST-JANAF Thermochemical Tables.*[‡]

The phase diagram for Ag and Cu derived with this method is shown in Fig. 7.3. This phase diagram spans a wide range of O_2 pressures, from those relevant in ultra-high-vacuum experiments to ambient pressures. The diagram confirms what you probably already knew from experience with coins and jewelry, namely that Ag is much more resistant to oxidation than Cu. It is important to remember that this phase diagram only describes the equilibrium state of each material; it cannot give any information about the rates at which equilibrium is reached. At room temperature and ambient pressure, Cu converts to Cu_2O very slowly because large energy barriers exist for the processes that allow gas-phase O_2 to oxidize Cu.

It should not be surprising that the result from Fig. 7.3 is not exact. The temperature at which Ag and Ag_2O are equally stable at $p = 1$ atm is found experimentally to be $\sim$460 K, while our calculation predicted this temperature to be 655 K. This deviation occurs because of both approximations in our thermodynamic calculations and the inexactness of DFT. Our calculations

---

[†]For O_2 it is important that this calculation allows for spin polarization since the ground state of O_2 is paramagnetic.

[‡]The ideal-gas data in these tables were calculated directly from a statistical mechanical description of the isolated molecules. In terms of the quantities defined in these tables, $\tilde{\mu}_{O_2}(T, p^o) = [H^o(T) - H^o(T_r)] - TS(T) - [H^o(0) - H^o(T_r)]$. This expression is derived by noting that the chemical potential of an ideal gas is equal to the ideal-gas free energy, $G = H - TS$.

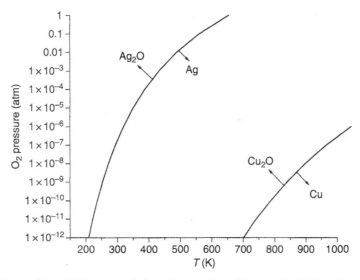

Figure 7.3 DFT-computed phase diagram for oxidation of bulk Cu and Ag.

neglected the influence of vibrational contributions to the free energies of the solids; as mentioned briefly above, these effects can be included at the price of performing considerably more involved calculations. More importantly, however, the total energies from our DFT calculations unavoidably include systematic errors because of the inexact nature of DFT.

7.1.1 Examples Including Disorder—Configurational Entropy

Each of the materials we looked at above is an ordered crystal, that is, there is only one way to arrange the set of atoms we are interested in to give the crystal structure. This is of course not true for all materials. Suppose we wanted to add to our class of possible materials a sample of Cu_2O with a small number of O vacancies. Specifically, imagine we used a sample of Cu_2O containing $2n$ Cu atoms and with k O atoms removed. There are

$$N_{config} = \frac{n!}{(n-k)!k!} \tag{7.7}$$

distinct ways these O vacancies could be arranged in the sample. The entropy associated with these different configurations follows directly from the definition of entropy:

$$S_{config} = k_B \ln N_{config}. \tag{7.8}$$

This quantity is called the configurational entropy. Using Stirling's approximation, this entropy can be written as

$$S_{\text{config}} = -k_B n[(1 - c)\ln(1 - c) + c \ln c], \qquad (7.9)$$

where $c = k/n$ is the concentration of O vacancies.

When we developed expressions for the grand potential above [Eqs. (7.1)–(7.5)], we simplified the free energy of each system by assuming that the entropic contributions to each material are similar. This assumption is clearly not reasonable if we are comparing ordered materials (e.g., Cu_2O with no vacancies) to disordered materials (e.g., Cu_2O with vacancies) because the configurational entropy of the two materials is quite different. To include this fact in our analysis of phase stability, Eq. (7.5) should be replaced by

$$\Omega_i(T, \mu_O) = E_i - TS_{\text{config},i} - \mu_O N_{O,i} - \Omega_S^M. \qquad (7.10)$$

Here, the last term is constant for all materials with the same number of Cu atoms within the approximation that entropic contributions from atomic vibrations are equivalent in each material.

When we compared Cu and Cu_2O, we used the grand potential to determine which of these materials was the most stable. We can now tackle a slightly different question, namely, what concentration of O vacancies should exist in Cu_2O when this material is in equilibrium with gaseous O_2 at some specified pressure and temperature? To answer this question, we perform two DFT calculations. First, we calculate the energy of vacancy-free Cu_2O using a supercell containing $2n$ Cu atoms and n O atoms, E, and express this energy as $E = nE_{\text{crystal}}$, where E_{crystal} is the energy per formula unit of the stoichiometric material. Second, we remove one O atom from the supercell and recompute the energy. If the supercell we used in these calculations is large enough that the vacancies can be thought of as isolated from their periodic images, then we can write the total energy from this second calculation as $E = (n - 1)E_{\text{crystal}} + E_{\text{vac}}$, where E_{vac} is the energy of a formula unit containing a vacancy that is surrounded by the stoichiometric material. We can view this expression as a definition for E_{vac}, as long as we remember that it is only an accurate definition if our supercell is large enough. This calculation for a supercell containing a vacancy should be performed while holding the lattice constants for the supercell equal to those of the stoichiometric material since we are interested in a situation where a very low concentration of vacancies causes local distortions in the atomic positions but does not change the material's overall crystal structure.

Once we have determined $E_{crystal}$ and E_{vac} in this way, we can write the energy for a sample of the material with N formula units and m vacancies as

$$\frac{E(N, m)}{N} \cong (1 - c)E_{crystal} + cE_{vac}$$

where $c = m/N$. If we use this energy in the grand potential, Eq. (7.10), we have a definition of the grand potential in terms of the vacancy concentration. As before, we want to find the material that minimizes the grand potential, which can be done by finding the value of c for which $\partial\Omega/\partial c = 0$. When $c \ll 1$, this condition leads to a simple result:

$$c = \exp[-\beta(E_{vac} - E_{crystal} + \mu_O)]. \qquad (7.11)$$

Figure 7.4 shows the vacancy concentrations computed in this way for three different O_2 pressures. Because of the dependence of the oxygen chemical potential on pressure [see Eq. (7.6)], increasing the oxygen pressure dramatically reduces the vacancy concentration. The results in Fig. 7.4 justify two approximations that were made in our calculations above. First, the vacancy concentrations under physically relevant conditions are always very small, so determining the properties of a vacancy in a DFT calculation representing

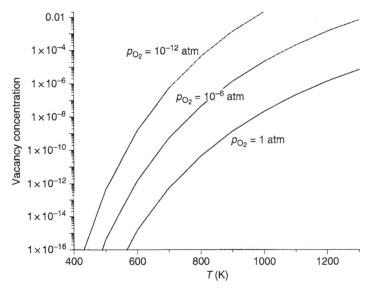

Figure 7.4 Oxygen vacancy concentration in Cu_2O calculated as described in the text at three different oxygen pressures. The dashed lines for the two lower pressures indicate the temperature region where Cu_2O is thermodynamically unfavored relative to Cu as defined in the phase diagram in Fig. 7.3.

a vacancy surrounded by a large region of vacancy-free material was appropriate. Second, at the conditions on Fig. 7.3 where the pure metal and metal oxide were found to be equally stable, the vacancy concentration in the oxide is again very low. This observation implies that constructing the phase diagram in Fig. 7.3 based on calculations that neglected the presence of any vacancies was reasonable.

7.2 STABILITY OF METAL AND METAL OXIDE SURFACES

The phase diagram for the oxidation of Ag that we examined in Section 7.1 illustrated the ideas of *ab initio* thermodynamics with a material that has properties that could be readily determined experimentally. Now we turn to a related topic where using experimental measurements to establish a phase diagram is more challenging—understanding the stability of metal and metal oxide surfaces. To motivate this result, we note that Ag is a well-known commercial catalyst for epoxidation of ethylene,[§] where the reaction takes place with an O_2 pressure of approximately atmospheric pressure and temperatures of 200–300°C. The bulk phase diagram of Ag indicates that the stable phase of the catalyst under these conditions (and under lower pressures of O_2) is metallic Ag. If the reaction is attempted using clean Ag under ultra-high-vacuum conditions, however, the surface is found to be effectively inactive as a catalyst.[1] This hints that the surface of the active Ag catalyst under industrial conditions is not metallic Ag. What then is the nature of the active catalyst surface?

The possibility that the real surface of an Ag catalyst involves some level of adsorbed oxygen can be readily explored using *ab initio* thermodynamics. If a catalog of candidate surface structures is created and the energy of each structure is calculated with DFT, then a phase diagram can be predicted in the same way as the bulk phase diagrams in Section 7.1. An example of a phase diagram for oxygen interacting with the Ag(111) surface calculated in this way by Li, Stampfl, and Scheffler[1] is shown in Fig. 7.5. This diagram includes results for Ag(111) with O atoms on the surface at coverages of less than one monolayer (ML) that are similar to the overlayers we examined in Chapter 4. These calculations also considered several more complicated structures in which O atoms are present not only on the surface but also below the surface atoms. Two of these structures are shown in Fig. 7.6. The structure with multiple layers of O atoms was not found to appear on the final phase diagram at all; this surface structure is not favored at any temperature/pressure combination (a result that is qualitatively similar to the situation with M_3O_2 in

[§]Over 9 million pounds per year of ethylene oxide are produced by this process in the United States alone.

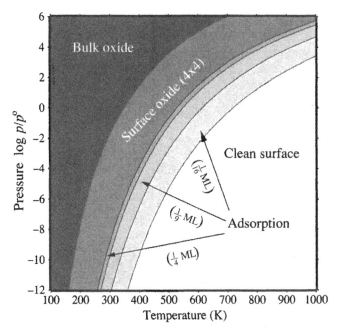

Figure 7.5 An *ab initio* phase diagram for O_2 interacting with Ag(111). [Reprinted with permission from W.-X. Li, C. Stampfl, and M. Scheffler, Insights into the Function of Silver as an Oxidation Catalyst by *ab initio* Thermodynamics, *Phys. Rev. B* **68** (2003), 165412 (Copyright 2003 by the American Physical Society).]

Fig. 7.2). The second structure, however, was found to play an important role in the overall phase diagram. This structure is called a surface oxide since the outermost layers of the material are in an oxide form while the bulk of the material is a pure metal. It can be seen from Fig. 7.5 that at most temperatures, there is a range of pressures spanning several orders of magnitude for which this surface oxide structure is more stable than either the bulk oxide or the clean metal surface. This phase diagram strongly suggests that the working catalyst under industrial conditions is a surface oxide rather than bare Ag.

The example above illustrates how constructing a phase diagram is relatively straightforward *once a list of candidate structures has been specified*. At the same time, the complexity of the surface oxide structure in Fig. 7.6 is an excellent example of why generating the relevant candidate structures is often far from straightforward. The structure shown in Fig. 7.6 was based on the best experimental data available on this ordered surface phase that were available at the time of Li, Stampfl, and Scheffler's[1] calculations. Since then, however, additional experiments and DFT calculations have indicated that the structure of the "true" surface oxide is somewhat different than the one shown in Fig. 7.6 and, moreover, other surface oxide phases with similar stabilities also exist.[2]

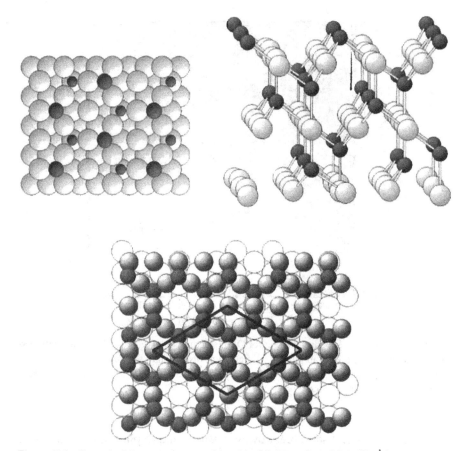

Figure 7.6   Several of the structures considered by Li, Stampfl, and Scheffler[1] in constructing their phase diagram for $O_2/Ag(111)$. The top left panel shows a top view of a structure with both surface and subsurface O atoms (large and small dark spheres, respectively). The top right panel shows a side view of the same structure. The DFT calculations do not predict this structure to be thermodynamically favored. The bottom panel shows a top view of the (4×4) surface oxide, which has a complex arrangement of Ag atoms (light spheres) and O atoms (dark spheres) on top of an Ag(111) substrate (unfilled circles). DFT calculations predict this structure to be favored at certain temperatures and pressures. (Reprinted by permission from the source cited in Fig. 7.5.)

7.3   MULTIPLE CHEMICAL POTENTIALS AND COUPLED CHEMICAL REACTIONS

In all of the examples above, the phase diagram of interest has been a function of temperature and the pressure of a single gas species, O_2. There are clearly many situations where phase diagrams involving the chemical potential of more than one species are of interest. To give just one example, removing sulfur from fuels is typically achieved using hydrodesulfurization (HDS), a process in which excess H_2 is reacted with fuels over a sulfide catalyst, forming H_2S. The structure of the catalyst for this reaction is a function of both the hydrogen chemical potential, μ_H and the sulfur chemical potential, μ_S. Although it is beyond the scope of our presentation here, the methods we have already introduced in this chapter define the basis for developing phase diagrams for situations such as this one based on DFT calculations of candidate material structures. The Further Reading section at the end of the chapter lists several examples of calculations of this kind.

A final caveat that must be applied to phase diagrams determined using DFT calculations (or any other method) is that not all physically interesting phenomena occur at equilibrium. In situations where chemical reactions occur in an open system, as is the case in practical applications of catalysis, it is possible to have systems that are at steady state but are not at thermodynamic equilibrium. To perform any detailed analysis of this kind of situation, information must be collected on the rates of the microscopic processes that control the system. The Further Reading section gives a recent example of combining DFT calculations and kinetic Monte Carlo calculations to tackle this issue.

EXERCISES

1. Lithium hydride, LiH, decomposes (under appropriate conditions of temperature and pressure) to solid Li and gaseous H_2. Solid Li is a bcc crystal, while LiH has the NaCl structure. Construct an *ab initio* phase diagram for LiH and Li as a function of temperature and H_2 pressure.

2. Extend your calculations for LiH to estimate the H vacancy concentration in LiH. Evaluate this concentration at room temperature and $P_{H_2} = 1$ atm; then repeat your calculation for conditions that might be possible in ultra-high-vacuum, $P_{H_2} = 10^{-9}$ atm.

3. Derive Eq. (7.11).

4. Construct a surface phase diagram for H atoms adsorbed on Cu(100) as a function of temperature and H_2 pressure by comparing the relative stability of ordered overlayers of H with coverages of 1, $\frac{1}{2}$, $\frac{1}{4}$, and $\frac{1}{9}$ of a monolayer. You could also consider situations where H atoms are present in the interstitial sites underneath the topmost layer of surface atoms.

5. Extend the ideas from Section 7.1.1 to estimate the concentration of interstitial H atoms in the octahedral sites of bulk Cu for the same conditions you looked at in the previous exercise.

REFERENCES

1. W.-X. Li, C. Stampfl, and M. Scheffler, Insights into the Function of Silver as an Oxidation Catalyst by *ab initio* Thermodynamics, *Phys. Rev. B* **68** (2003), 165412, gives detailed references for the years of experimental work that went into understanding this problem.
2. J. Schnadt et al., Revisiting the Structure of the *p*(4 × 4) Surface Oxide on Ag(111), *Phys. Rev. Lett.* **96** (2006), 146101.

FURTHER READING

The statistical thermodynamics that underlies the results in this chapter are described in most textbooks on molecular statistical mechanics. One well-known example is D. A. McQuarrie, *Statistical Mechanics*, University Science Books, Sausalito, CA, 2000.

The *NIST-JANAF Thermodynamics Tables* are available online from the National Institute of Standards: www.nist.gov/srd/monogr.htm.

Our discussion of vacancies in Section 7.1.1 only looked at the simplest situation where a vacancy is created by removing one atom and all its electrons. In some insulating materials, charged defects are more stable than uncharged defects. A good example of using DFT to describe this situation is:

C. G. van de Walle and J. Neugabauer, Universal Alignment of Hydrogen Levels in Semiconductors, Insulators, and Solutions, *Nature* **423** (2003), 626.

For other examples where a full description of vacancies is more complicated than our discussion, see the following:

V. Ozolins, B. Sadigh, and M. Asta, Effects of Vibrational Entropy on the Al-Si Phase Diagram, *J. Phys.: Condens. Matt.* **17** (2005), 2197.

G. Lu and E. Kaxiras, Hydrogen Embrittlement in Aluminum: The Crucial Role of Vacancies, *Phys. Rev. Lett.* **94** (2005), 155501.

C. J. Forst, J. Slycke, K. J. Van Vliet, and S. Yip, Point Defect Concentrations in Metastable Fe-C Alloys, *Phys. Rev. Lett.* **96** (2006), 175501.

For examples of phase diagrams that involve multiple chemical potentials, see the following:

S. Cristol, J. F. Paul, E. Payen, D. Bougeard, S. Clemendot, and F. Hutschka, Theoretical Study of the MoS_2 (100) Surface: A Chemical Potential Analysis of Sulfur and Hydrogen Coverage. 2. Effect of the Total Pressure on Surface Stability, *J. Phys. Chem. B* **106** (2002), 5659.

C. Stampfl and A. J. Freeman, Stable and Metastable Structures of the Multiphase Tantalum Nitride System, *Phys. Rev. B* **71** (2005), 024111.

The second of these articles is an example where the chemical potential of one of the relevant species is not defined with respect to a gaseous state.

To learn more about situations where coupled chemical reactions can influence the observed state of a system, see the following:

K. Reuter and M. Scheffler, First-Principles Kinetic Monte Carlo Simulations for Heterogeneous Catalysis: Application to the CO Oxidation at $RuO_2(110)$, *Phys. Rev. B* **73** (2006), 045433.

APPENDIX   CALCULATION DETAILS

All calculations in Section 7.1 used the PW91 GGA functional. Bulk Ag and Cu were treated with cubic supercells containing four atoms, while the cubic supercells for bulk Ag_2O and Cu_2O contained six atoms. For each bulk material, reciprocal space was sampled with $10 \times 10 \times 10$ k points placed with the Monkhorst–Pack method. Calculations for O_2 used a $10 \times 10 \times 10$ Å supercell containing one molecule and $3 \times 3 \times 3$ k points. Spin polarization was used in the calculations for O_2, but spin has no effect on the bulk materials considered. The energy cutoff for all calculations was 396 eV.

The calculations in Section 7.1.1 examining vacancies in Cu_2O used a cubic supercell containing 16 formula units of the compound and $5 \times 5 \times 5$ k points. All other details of these calculations were the same as those for the bulk materials in Section 7.1.

8

ELECTRONIC STRUCTURE AND MAGNETIC PROPERTIES

In previous chapters we focused on physical properties for which electrons are only important in the sense that we must know the ground state of the electrons to understand the material's energy. There is, of course, a long list of physical properties where the details of the electronic structure in the material are of great importance. Two examples of these properties are the classification of a bulk material as a metal, a semiconductor, or an insulator and the existence and characteristics of magnetic properties. In this chapter we examine how information related to these questions can be obtained from DFT calculations.

8.1 ELECTRONIC DENSITY OF STATES

One of the primary quantities used to describe the electronic state of a material is the electronic density of states (DOS):

$\rho(E) \, dE$ = number of electron states with energies in interval $(E, E + dE)$.

Recall from Chapter 3 that the basic idea of a plane-wave DFT calculation is to express the electron density in functions of the form $\exp(i\mathbf{k} \cdot \mathbf{r})$. Electrons associated with plane waves of this form have energy $E = (\hbar\mathbf{k})^2/2\,m$. As a result, once a DFT calculation has been performed, the electronic DOS can

Density Functional Theory: A Practical Introduction. By David S. Sholl and Janice A. Steckel
Copyright © 2009 John Wiley & Sons, Inc.

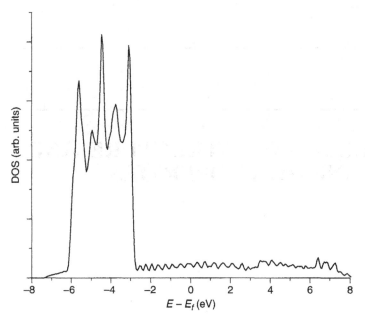

Figure 8.1 Electronic DOS for bulk Ag, calculated using a supercell containing two atoms and sampling k space with $24 \times 24 \times 24$ k points.

be determined by integrating the resulting electron density in k space. Below we give a series of examples to illustrate calculations of this type. Before reading through these examples, it would be a good idea to briefly review the basic concepts associated with k space in Section 3.1 if you do not feel completely confident with them.

Our first example, the DOS for bulk Ag, is shown in Fig. 8.1. There are several points to notice associated with the numerical details of this calculation. First, a large number of k points were used, at least relative to the number of k points that would be needed to accurately determine the total energy of Ag (see, e.g., Fig. 3.2). Using a large number of k points to calculate the DOS is necessary because, as described above, the details of the DOS come from integrals in k space.

A second point is that we plotted the energy in Fig. 8.1 not in terms of an absolute energy but instead relative to the Fermi energy, E_f. The Fermi energy is the energy of the highest occupied electronic state.* A simple observation from Fig. 8.1 is that the DOS for Ag is nonzero at the Fermi energy. The most straightforward definition of a metal is that metals are materials with a nonzero DOS at the Fermi level. One useful way to think about this is

---

*More precisely, the highest occupied state at $T = 0$ K, since at nonzero temperatures thermal excitations of electrons lead to some population of states above the Fermi energy.

to consider what happens when an electric field is applied to the material. An electric field will accelerate electrons to higher energies than the electrons have when there is no field. In a metal, there are electronic states just above the Fermi energy that can be populated by these accelerated electrons, and as a result the material can readily conduct electricity. The result in Fig. 8.1 agrees with what you already knew about Ag—it is a metal.

The third numerical detail that is important in Fig. 8.1 arises from the fact that Ag is a metal and the observation that the DOS is obtained from integrals in k space. In Section 3.1.4 we described why performing integrals in k space for metals holds some special numerical challenges. The results in Fig. 8.1 were obtained by using the Methfessel and Paxton smearing method to improve the numerical precision of integration in k space.

Figure 8.2 shows a similar calculation of the electronic DOS for bulk Pt. As for Ag, the Pt DOS is nonzero at the Fermi energy, indicating that Pt is a metal. Two qualitative observations that can be made in comparing the DOS of Ag and Pt are (i) Ag has its electronic states concentrated in a smaller range of energies than Pt and (ii) Ag has a lower density of unoccupied states above the Fermi energy than Pt. Both of these observations can be loosely correlated with the physical observation that Pt is typically much more chemically reactive than Ag.

Our next example is a different kind of material: silicon. Figure 8.3 shows the calculated DOS for bulk Si from a DFT calculation using a two-atom

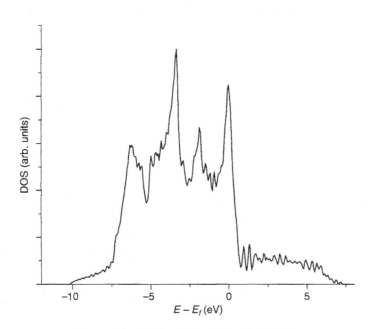

Figure 8.2 Similar to Fig. 8.1 but for bulk Pt.

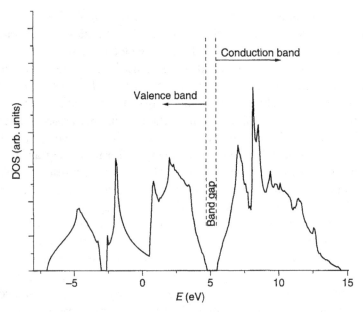

Figure 8.3   Calculated electronic DOS for bulk Si.

supercell and $12 \times 12 \times 12$ k points. As indicated on the figure, the DOS can be divided into two separate regions, the valence band and the conduction band. The valence band is the collection of all occupied electronic states, while all states in the conduction band are unoccupied (at $T = 0$ K). The region of energy that separates the valence and conduction bands contains no electronic states at all; this is the band gap. The qualitative description of electron conduction in a metal above can also be applied to materials with a band gap, and it is clear that applying an electric field to these materials will not lead to electron conduction as easily as it does in metals. Materials with a band gap are classified as either semiconductors if their band gap is "small" or insulators if their band gap is "wide." The distinction between these two types of materials is somewhat arbitrary, but band gaps larger than $\sim$3 eV are typically considered wide band gaps.

It is important to note that while the DFT calculation in Fig. 8.3 correctly predicts the existence of a band gap for Si, the width of this gap is not predicted accurately by DFT. Our calculation predicts a band gap of 0.67 eV, but the experimentally observed band gap for Si is 1.1 eV.[†]

We commented above that special techniques must be applied to accurately perform integrals in k space for metals. It is important to note that the smearing

[†]The underestimation of band gaps is a systematic difficulty with DFT. Much work has been devoted to corrections to improve the accuracy of predictions for band gaps (see Further Reading).

methods that are used for metals should not be used for semiconducting and insulating materials. The Si calculations in Fig. 8.3 were performed using Blöchl's tetrahedron method (see Section 3.1.4) for interpolation in k space but no smearing.

A final observation from Fig. 8.3 is that there are a number of places where the slope of the DOS changes discontinuously. These are known as van Hove singularities.

The properties of the band gap in semiconductors often control the applicability of these materials in practical applications. To give just one example, Si is of great importance as a material for solar cells. The basic phenomenon that allows Si to be used in this way is that a photon can excite an electron in Si from the valence band into the conduction band. The unoccupied state created in the valence band is known as a hole, so this process has created an electron-hole pair. If the electron and hole can be physically separated, then they can create net electrical current. If, on the other hand, the electron and hole recombine before they are separated, no current will flow. One effect that can increase this recombination rate is the presence of metal impurities within a Si solar cell. This effect is illustrated in Fig. 8.4, which compares the DOS of bulk Si with the DOS of a large supercell of Si containing a single Au atom impurity. In the latter supercell, one Si atom in the pure material was replaced with a Au atom,

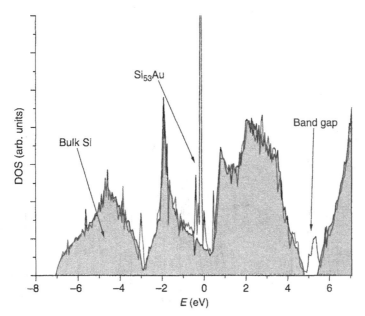

Figure 8.4 Electronic DOS of bulk Si (black line) and bulk $Si_{53}Au$ (grey line). Both DFT calculations were performed with a 54-atom supercell and sampled k space with $12 \times 12 \times 12$ k points.

and the positions of all atoms in the supercell were relaxed until the system's energy was minimized. The lattice constant of the supercell was held fixed at the value for bulk Si, however, since the very low density of metal impurities that would be present in a real material would not appreciably change the Si lattice constant.

The most important feature in Fig. 8.4 is the existence of states in the Au-doped Si within the band gap of pure Si. These new states make the recombination of electron–hole pairs far more rapid in the doped material than in pure Si. That is, the existence of Au impurities in crystalline Si has serious negative effects on the properties of the material as a solar cell. This problem does not only occur with Au; a large number of metal impurities are known to cause similar problems. More information about this topic is available from the sources in the Further Reading section at the end of the chapter.

To give an example of a wide band gap material, Fig. 8.5 shows the calculated DOS of bulk quartz (SiO_2). The band gap from the calculation shown in this figure is 5.9 eV. Even remembering that DFT typically underestimates the band gap, it is clear that quartz is an insulator. Unlike Si, the valence band of quartz includes several separate energy bands with distinct energy gaps between them. We will return to these bands when we discuss the local density of states in Section 8.2.

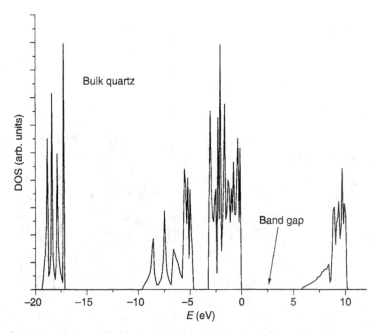

Figure 8.5 Electronic DOS of bulk quartz from DFT calculations with nine-atom supercell that sampled k space with $9 \times 9 \times 9$ k points. Energy is defined so the top of the valence band is at 0 eV.

As a final example of the DOS of a bulk material, we look at one of the oxides we examined in Chapter 7, Ag_2O. The DOS calculated using a GGA functional is shown in Fig. 8.6, and it shows a small but nonzero population of states above the Fermi energy, predicting that Ag_2O is a metal. This prediction is not sensitive to the numerical details of the calculation; the same prediction is obtained with or without smearing in k space, and with LDA or GGA functionals. Unfortunately, this prediction does not agree with experimental observations, which indicate that Ag_2O is a semiconductor with a band gap of ~1.3 eV. In this case, standard DFT calculations fail to correctly describe the electronic states near the Fermi energy. Similar failures occur for a number of other narrow band-gap semiconductors, including Ge, InN, and ScN. It is important to note that the failure of DFT to describe the semiconducting character of Ag_2O does not mean that DFT fails for all physical properties of this material. The DFT calculations associated with Fig. 8.6 predict O–Ag and Ag–Ag distances of 2.10 and 3.43 Å; these are in reasonable agreement with the experimental values of 2.04–2.05 and 3.34–3.35 Å, respectively.

The electronic DOS condenses the properties of the electronic states for all possible positions in reciprocal space into a simple form. A more nuanced view of a material's electronic structure is often possible by examining the material's band structure. The band structure represents the energy of the

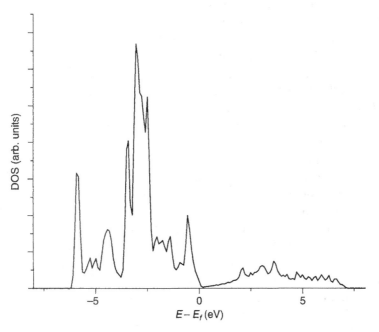

Figure 8.6   The DOS of Ag_2O calculated with DFT using a six-atom supercell and $20 \times 20 \times 20$ k points.

available electronic states along a series of lines in reciprocal space that typically form a closed loop beginning and ending at the Γ point. Band structure diagrams can be routinely calculated from plane-wave DFT calculations. Attention must be paid in calculations of this kind to the placement of k points, as the electronic states must be evaluated as a series of k points spaced close together along the special directions in reciprocal space relevant for the band structure diagram.

8.2   LOCAL DENSITY OF STATES AND ATOMIC CHARGES

To interpret the electronic structure of a material, it is often useful to understand what states are important in the vicinity of specific atoms. One standard way to do this is to use the local density of states (LDOS), defined as the number of electronic states at a specified energy weighted by the fraction of the total electron density for those states that appears in a specified volume around a nuclei. Typically, this volume is simply taken to be spherical; so to calculate the LDOS we must specify the effective radii of each atom of interest. This definition cannot be made unambiguously. If a radius that is too small is used, information on electronic states that are genuinely associated with the nuclei will be missed. If the radius is too large, on the other hand, the LDOS will include contributions from other atoms.

Figures 8.7 and 8.8 show the local DOS for O and Si atoms in bulk quartz. In these calculations, the radii of O and Si were set to 1.09 and 1.0 Å, respectively. The total DOS for this material was shown in Fig. 8.5. Each LDOS is split into contributions from s and p bands in the electronic structure. It can be seen that the band at the top of the valence band is dominated by p states from O atoms, with a small contribution from Si p states. The band located 5–10 eV below the valence band edge is a mixture of O atom p states and Si atom s and p states. The band furthest below the valence band edge is a mixture of Si and O s states.

It is often convenient to think of atoms within bulk materials or molecules as having net charges. In an ionic material such as NaCl, for example, it is conventional to associate charges of $+1$ and -1 (in units of electron charge) to Na and Cl atoms. The ambiguity in defining the volumes used for LDOS calculations illustrates why making this assignment from a calculated electron density is not necessarily a simple task. This is a longstanding challenge in computational chemistry in general, not just for plane-wave DFT calculations, and many different methods have been explored as possible solutions. Many of the most widely used methods, such as Mulliken analysis and the ChelpG method, rely on having a well-defined basis set of local functions, so they are not typically available within plane-wave DFT calculations.

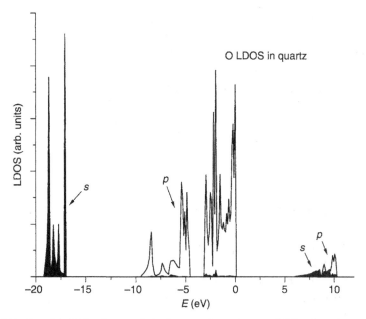

Figure 8.7 Local DOS for O atoms in bulk quartz from the same DFT calculations described in Fig. 8.5 using a radius of 1.09 Å for O.

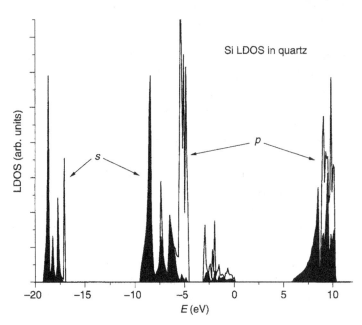

Figure 8.8 Local DOS for Si atoms in bulk quartz from the same DFT calculations described in Fig. 8.5 using a radius of 1.0 Å for Si.

An elegant method that is applicable within plane-wave calculations is the Bader decomposition, which uses stationary points in the three-dimensional electron density to partition electrons among different atoms. It is important to note that the various methods we have just mentioned do *not* necessarily give the same result when they are used in quantum chemistry methods where they are all applicable. As a result, in any calculations that assign atomic charges it is important to report the method that was used to make this assignment and to keep the ambiguity of these methods in mind when interpreting the results.

8.3 MAGNETISM

Magnetism is a direct consequence of the nonzero spin of electrons. In diamagnetic materials, each electronic state is populated by two electrons, one with spin up and another with spin down. This situation is shown schematically for a periodic material in Fig. 8.9a. In magnetic materials, however, electronic states exist that contain only one electron. When unpaired electrons exist, they can be ordered in many different ways, each defining a different magnetic state. The two most common magnetic states are shown in Figs. 8.9b and c; these are the ferromagnetic state with all electron spins pointing in the same direction and the antiferromagnetic state with electron spins alternating in direction on adjacent atoms. More subtle spin orderings are also possible, as illustrated by one simple example in Fig. 8.9d. The average electron spin per atom is known as the magnetic moment. The ferromagnetic state shown in Fig. 8.9b has a magnetic moment of 1, while all the other examples in the same figure have zero magnetic moment.

It is relatively common for DFT calculations to not explicitly include electron spin, for the simple reason that this approximation makes calculations faster. In materials where spin effects may be important, however, it is crucial that spin is included. Fe, for example, is a metal that is well known for its magnetic properties.[‡] Figure 8.10 shows the energy of bulk Fe in the bcc crystal structure from calculations with no spin polarization and calculations with ferromagnetic spin ordering. The difference is striking; electron spins lower the energy substantially and increase the predicted equilibrium lattice constant by ~ 0.1 Å.

Figure 8.9 illustrates an important subtlety in including spin in DFT calculations, namely that many distinct magnetic orderings of electron spins are possible. The optimization of electron spins in a single DFT calculation is

---

[‡]Compasses using the magnetic properties of iron have been used for around 1000 years.

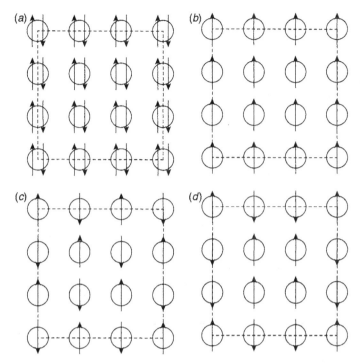

Figure 8.9 Schematic illustrations of spin states in a two-dimensional periodic material. Circles indicate individual atoms and the dotted lines show a single supercell. In (*a*), all electrons are paired on each atom. In the remaining examples, a single unpaired electron exists on each atom. Examples of a ferromagnetic state, an antiferromagnetic state and a more complex magnetic state are shown in (*b*), (*c*), and (*d*), respectively.

similar to the optimization of the atom positions within a specified crystal structure; an initial approximation for the spins is made, and the calculation subsequently finds a local minimum associated with this initial approximation. Typically, a ferromagnetic state is used as an initial approximation unless otherwise specified. A key observation here is that this approach can only give a *local* minimum on the energy surface defined by all possible spin orderings. Determining the spin ordering associated with the *global* minimum on this energy surface is more difficult; this problem is completely analogous to the task of predicting the best crystal structure for a material from all possible crystal structures. As in the case of using DFT to explore crystal structures, it is wise to examine more than one initial approximation for the spin ordering when examining magnetic ordering. As an example, Fig. 8.10 shows the energy of fcc Fe with antiferromagnetic ordering from a series of calculations in which this ordering was used as the initial approximation for the spin states. The antiferromagnetic energies lie considerably higher in energy than the

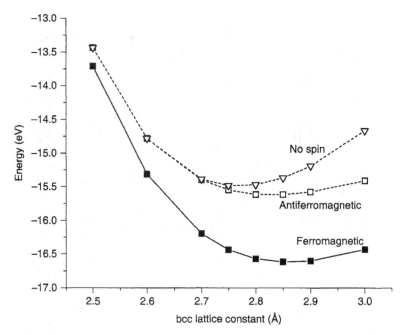

Figure 8.10   Energy of bcc Fe as a function of the Fe lattice constant from DFT calculations with several different spin states. The results labeled 'No spin' are from calculations without spin polarization.

ferromagnetic results, in agreement with the experimental fact that Fe is ferromagnetic. Note that these calculations do not provide any information about the possible existence of ordered spin states with structures more complex than simple ferromagnetic or antiferromagnetic ordering.

We must emphasize that in this extremely brief introduction to magnetic properties we have only scratched the surface of this wide-ranging topic. Because magnetic properties lie at the heart of most digital data storage techniques, the numerous phenomena associated with magnetism have been the subject of intense worldwide research for decades. The Further Reading section lists several resources that can provide an entry point into this topic.

EXERCISES

1. Calculate the electronic DOS and total energy for bulk Ag using different numbers of k points to sample k space. Do your results agree with the comment made in relation to Fig. 8.1 that more k points are needed to get well-converged results for the DOS than for the total energy? Explain in qualitative terms why this occurs.

2. Calculate the electronic DOS for bulk Ag_2O after optimizing the structure of this material. Use your calculations to check our claim that DFT calculations predict this material to be a metal.

3. Sketch three spin orderings analogous to the one shown in Fig. 8.9*d* but with nonzero net magnetic moment.

4. Co, MnBi, and MnF_2 are all magnetic materials. Determine whether they are ferromagnetic or antiferromagnetic. How strongly does the magnetic state of each material affect its lattice constant? MnBi forms the NiAs structure (space group $R\bar{3}$ m) and MnF_2 forms the rutile (TiO_2) structure.

5. Determine using calculations whether spin has any effect on the structure and energy of the diatomic molecules N_2, O_2, and CO.

6. The DFT calculations shown in Fig. 8.10 are valid for bulk Fe at $T = 0$ K. At temperatures above the so-called Curie temperature, T_c, Fe is no longer ferromagnetic because of thermal effects on the spin order. Find an appropriate reference that tells you what metals in the periodic table are ferromagnetic and what their Curie temperatures are.

FURTHER READING

For more details on the roles of metal impurities in silicon solar cells and efforts to engineer materials that are less affected by these impurities, see S. Dubois, O. Palais, M. Pasquinelli, S. Martinuzzi, and C. Jassaud, *J. Appl. Phys.* **100** (2006), 123502, and T. Buonassisi, A. A. Istratov, M. A. Marcus, B. Lai, Z. H. Cai, S. M. Heald, and E. R. Weber, *Nature Materials* **4** (2005), 676.

For an explanation of why DFT does not accurately predict band gaps, see J. P. Perdew and M. Levy, *Phys. Rev. Lett.* **51** (1983), 1884 and L. J. Sham and M. Schluter, *Phys. Rev. Lett.* **51** (1983), 1888. For a description of some correction methods, see S. Lany and A. Zunger, *Phys. Rev. B* **78** (2008), 235104.

For a detailed discussion of the failure of DFT calculations to predict the semiconducting character of Ag_2O and the application of "post-GGA" methods to overcome this problem, see W.-X. Li, C. Stampfl, and M. Scheffler, *Phys. Rev. B* **68** (2003), 165412.

To learn more about charge decomposition methods for assigning atomic charges, see G. Henkelman, A. Arnaldsson, and H. Jonsson, *Comp. Mater. Sci.* **36** (2006), 354.

The basic principles of magnetism are described in essentially every solid-state physics text. A classic reference for this genre is C. Kittel, *Introduction to Solid State Physics*, Wiley, New York, 1976. Several books that focus more specifically on magnetic phenomena are M. Getzlaff, *Fundamentals of Magnetism*, Springer,

Berlin, 2007, P. Mohn, *Magnetism in the Solid State: An Introduction*, Springer, Berlin, 2005, and A. H. Morrish, *The Physical Principles of Magnetism*, Wiley– IEEE Press, New York, 2001.

APPENDIX   CALCULATION DETAILS

All results in this chapter were obtained from calculations with the PW91 GGA functional. The Monkhorst–Pack method for sampling k space was used in all cases.

Section 8.1   Calculations for bulk Ag, Pt, Si, quartz, and Ag_2O were performed using supercells with the DFT-optimized lattice parameter and energy cutoffs of 396, 239, 151, 396, and 396 eV, respectively. The supercell size and number of k points for Ag, Pt, $Si_{53}Au$, quartz, and Ag_2O were given above. The bulk Si results in Fig. 8.3 are from calculations using a 2-atom supercell and $12 \times 12 \times 12$ k points. The results in Fig. 8.4 are from calculations with a 54-atom supercell using an energy cutoff of 181 eV.

Section 8.3   All calculations for bcc Fe were performed with a cubic supercell containing two atoms, $5 \times 5 \times 5$ k points, and an energy cutoff of 270 eV.

9

AB INITIO MOLECULAR DYNAMICS

An unavoidable characteristic of the materials that surround us is that their atoms are in constant motion. There are a large set of topics for which knowing how the atoms in a material move as a function of time is a prerequisite for describing some property of practical interest. In this chapter, we explore the methods of molecular dynamics (MD), a set of computational tools that makes it possible to follow the trajectories of moving atoms. We begin in Section 9.1 by stepping back from quantum mechanics temporarily and describing how MD simulations are performed using classical mechanics. This detour introduces several concepts that are critical in using MD within DFT calculations. Section 9.2 introduces the methods that can be used to perform MD simulations using information derived from DFT calculations. Section 9.3 gives two examples of using *ab initio* MD, one involving the generation of disordered bulk materials and another in which MD is used to explore complicated energy surfaces.

9.1 CLASSICAL MOLECULAR DYNAMICS

9.1.1 Molecular Dynamics with Constant Energy

To understand the main ideas that define *ab initio* MD, it is useful to first review some concepts from classical mechanics. Classical MD is a well-developed approach that is widely used in many types of computational chemistry and

Density Functional Theory: A Practical Introduction. By David S. Sholl and Janice A. Steckel
Copyright © 2009 John Wiley & Sons, Inc.

materials modeling. The Further Reading section at the end of the chapter includes resources that give a detailed entry point into this area.

We will consider a situation where we have N atoms inside a volume V, and we are interested in understanding the dynamics of these atoms. To specify the configuration of the atoms at any moment in time, we need to specify $3N$ positions, $\{r_1, \ldots, r_{3N}\}$, and $3N$ velocities, $\{v_1, \ldots, v_{3N}\}$. Two quantities that are useful for describing the overall state of our system are the total kinetic energy,

$$K = \frac{1}{2} \sum_{i=1}^{3N} m_i v_i^2 \qquad (9.1)$$

where m_i is the mass of the atom associated with the ith coordinate, and the total potential energy,

$$U = U(r_1, \ldots, r_{3N}). \qquad (9.2)$$

Newton's laws of motion apply to these atoms since we are treating their motion within the framework of classical mechanics. That is,

$$F_i = m_i a_i = m_i \frac{dv_i}{dt}, \qquad (9.3)$$

where F_i and a_i are the force and acceleration, respectively, acting on the ith coordinate and t is time. This force is also related to the derivative of the total potential energy by

$$F_i = -\frac{\partial U}{\partial r_i}. \qquad (9.4)$$

These relationships define the equations of motion of the atoms, which can be written as a system of $6N$ first-order ordinary differential equations:

$$\begin{aligned}
\frac{dr_i}{dt} &= v_i \\
\frac{dv_i}{dt} &= -\frac{1}{m_i} \frac{\partial U(r_1, \ldots, r_{3N})}{\partial r_i}.
\end{aligned} \qquad (9.5)$$

One important property of these equations is that they conserve energy; that is, $E = K + U$ does not change as time advances. In the language of statistical mechanics, the atoms move within a *microcanonical ensemble*, that is, a set of possible states with fixed values of N, V, and E. Energy is not the only conserved quantity in this ensemble. The total momentum of the particles in the system is also a conserved quantity. For an isolated collection of particles, the total angular momentum is also conserved, but this quantity is not conserved in systems with periodic boundary conditions.

If we are going to relate the properties of our system to a physical situation, we need to be able to characterize the system's temperature, T. In a macroscopic collection of atoms that is in equilibrium at temperature T, the velocities of the atoms are distributed according to the Maxwell–Boltzmann distribution. One of the key properties of this distribution is that the average kinetic energy of each degree of freedom is

$$\frac{1}{2}m\overline{(v^2)} = \frac{k_B T}{2}. \tag{9.6}$$

In molecular dynamics, this relationship is turned around to *define* temperature by

$$\frac{k_B T_{MD}}{2} \equiv \frac{1}{6N} \sum_{i=1}^{3N} m_i v_i^2. \tag{9.7}$$

Notice that because the kinetic energy, K, is not conserved by the equations of motion above, the temperature observed in a microcanonical ensemble MD simulation must fluctuate with time.

In essentially all systems that are physically interesting, the equations of motion above are far too complicated to solve in closed form. It is therefore important to be able to integrate these equations numerically in order to follow the dynamics of the atoms. A simple way to do this is to use the Taylor expansion:

$$r_i(t + \Delta t) = r_i(t) + \frac{dr_i(t)}{dt}\Delta t + \frac{1}{2}\frac{d^2 r_i(t)}{dt^2}\Delta t^2 + \frac{1}{6}\frac{d^3 r_i(t)}{dt^3}\Delta t^3 + O(\Delta t^4). \tag{9.8}$$

Rewriting this expansion by assigning the names of the derivatives on the right-hand side* gives

$$r_i(t + \Delta t) = r_i(t) + v_i(t)\Delta t + \frac{1}{2}a_i(t)\Delta t^2 + \frac{1}{6}\frac{d^3 r_i(t)}{dt^3}\Delta t^3 + O(\Delta t^4). \tag{9.9}$$

---

*The third derivative term (which is also the first time derivative of acceleration) also has a name: the jerk.

If you express Eq. (9.9) using a positive and a negative time step and take the difference of the two expressions, you can show that

$$r_i(t + \Delta t) \cong 2r_i(t) - r_i(t - \Delta t) + \frac{F_i(t)}{m_i} \Delta t^2. \qquad (9.10)$$

This is known as the Verlet algorithm. Provided that the time step, Δt, is sufficiently small, this algorithm gives an accurate approximation to the true trajectory defined by Eqs. (9.5).

We introduced the equations of motion above from the Newtonian point of view using Newton's famous equation $F = ma$. It is useful to realize that this is not the only (or even the best) way to define equations of motion within classical dynamics. Another powerful approach to this task is to define a quantity called the Lagrangian, L, in terms of the kinetic and potential energies,

$$L = K - U = \frac{1}{2} \sum_{i=1}^{3N} m_i v_i^2 - U(r_1, \ldots, r_{3N}). \qquad (9.11)$$

The equations of motion for each coordinate in terms of the Lagrangian are

$$\frac{d}{dt} \left(\frac{\partial L}{\partial v_i} \right) = \frac{\partial L}{\partial r_i}. \qquad (9.12)$$

You can verify in just a few lines that this approach defines the same equations of motion that we defined from the Newtonian approach above.

9.1.2   Molecular Dynamics in the Canonical Ensemble

In many applications of molecular dynamics, we would like to compare MD results with experimental observations. Under typical experimental conditions, the atoms of a material of interest are able to exchange heat with their surroundings. In this situation, the atoms exist in a *canonical ensemble*, where N, V, and T are constant. There are a number of ways to adapt the microcanonical MD method outlined above to mimic a canonical ensemble. One of the most elegant was originally introduced by Nosé, who began by using the Lagrangian for the microcanonical ensemble, Eq. (9.11), and forming an *extended Lagrangian*:

$$L = \frac{1}{2} \sum_{i=1}^{3N} m_i s^2 v_i^2 - U(r_1, \ldots, r_{3N}) + \frac{Q}{2} \left(\frac{ds}{dt} \right)^2 - gk_B T \ln s. \qquad (9.13)$$

Notice that if $s(t) \equiv 1$, then this extended Lagrangian reduces exactly to Eq. (9.11). Formally, we can determine the equations of motion associated with this extended Lagrangian using Eq. (9.12). These equations were written by Hoover in a convenient form using slightly different variables than the extended Lagrangian above:

$$\frac{dr_i}{dt} = v_i,$$

$$\frac{dv_i}{dt} = -\frac{1}{m_i}\frac{\partial U(r_1, \ldots, r_{3N})}{\partial r_i} - \frac{\xi}{m_i}v_i,$$

$$\frac{d\xi}{dt} = \frac{1}{Q}\left[\sum_{i=1}^{3N} m_i v_i^2 - 3Nk_B T\right],$$

$$\frac{d\ln s}{dt} = \xi.$$

$$(9.14)$$

These equations are not as hard to understand as they might appear at first glance. The first and second equations are the same as we saw in Eq. (9.5) apart from the addition of a "friction" term in the second equation that either increases or decreases the velocity, depending on the sign of ξ. The third equation controls the sign and magnitude of ξ. The meaning of this equation is clearer if we rewrite Eq. (9.7) as

$$\frac{d\xi}{dt} = \frac{3Nk_B}{Q}[T_{MD} - T].$$

$$(9.15)$$

This equation acts as a feedback control to hold the instantaneous temperature of the atoms in the simulation, T_{MD}, close to the desired temperature, T. If the instantaneous temperature is too high, ξ is smoothly adjusted by Eq. (9.15), leading to an adjustment in the velocities of all atoms that reduces their average kinetic energy. The parameter Q determines how rapidly the feedback between the temperature difference $T_{MD} - T$ is applied to ξ.

By applying the Taylor expansion as we did in Eq. (9.8), it is possible to derive an extension of the Verlet algorithm that allows these equations to be integrated numerically. This approach to controlling the temperature is known as the Nosé–Hoover thermostat.

9.1.3 Practical Aspects of Classical Molecular Dynamics

There are two aspects of classical MD calculations that are important to understand in order to grasp what kinds of problems can be treated with these methods. Both of these ideas are illustrated by examining the Verlet algorithm for constant energy MD, Eq. (9.10).

First, consider the computational effort involved in taking a single time step with this algorithm. During every time step, the force acting on each degree of freedom, $F_i(t)$, must be determined. This is (roughly) equivalent to calculating the total potential energy, $U = U(r_1, \ldots, r_{3N})$, at every time step. Even when this potential energy is a very simple function of the coordinates, for example, a sum of terms that depends only on the distance between atoms, the work involved in this calculation grows rapidly with system size. Because evaluating the forces dominates the computational cost of any MD simulation, great care must be taken to evaluate the forces in a numerically efficient way.

Second, the derivation of the Verlet algorithm from a Taylor expansion highlights the fact that this algorithm is only accurate for sufficiently small time steps, Δt. To estimate how small these time steps should be, we can note that in almost any atomic or molecular system, there is some characteristic vibration with a frequency on the order of 10^{13} s^{-1}. This means that one vibration takes ~100 fs. If we want our numerically generated MD trajectories to accurately describe motions of this type, we must take a large number of time steps in this period. This simple reasoning indicates that a typical MD simulation should use a time step no larger than ~10 fs. A trajectory following the dynamics of a set of atoms for just one nanosecond using a time step of 10 fs requires 10^5 MD steps, a significant investment of computational resources.

9.2   *AB INITIO* MOLECULAR DYNAMICS

The explanation of classical MD given above was meant in part to emphasize that the dynamics of atoms can be described provided that the potential energy of the atoms, $U = U(r_1, \ldots, r_{3N})$, is known as a function of the atomic coordinates. It has probably already occurred to you that a natural use of DFT calculations might be to perform molecular dynamics by calculating $U = U(r_1, \ldots, r_{3N})$ with DFT. That is, the potential energy of the system of interest can be calculated "on the fly" using quantum mechanics. This is the basic concept of *ab initio* MD. The Lagrangian for this approach can be written as

$$L = K - U = \frac{1}{2} \sum_{i=1}^{3N} m_i v_i^2 - E[\varphi(r_1, \ldots, r_{3N})], \qquad (9.16)$$

where $\varphi(r_1, \ldots, r_{3N})$ represents the full set of Kohn–Sham one-electron wave functions for the electronic ground state of the system. (See Section 1.4 for details.) This Lagrangian suggests that calculations be done in a sequential way: first, the ground-state energy is calculated; then the positions of the nuclei are advanced using one step of MD; then the new ground-state energy is calculated; and so on. We will refer to any method that advances the positions of nuclei along trajectories defined by classical mechanics from forces calculated from DFT as ab initio MD. Because of the limited size of the time steps that can be taken with MD, finding methods to perform these calculations with great efficiency was extremely important for making them feasible for physically interesting problems.

A key breakthrough that changed *ab initio* MD from simply an interesting idea to a powerful and useful method was made by Car and Parrinello (see Further Reading). They introduced an algorithm in which the separate tasks of following the motion of nuclei and finding the electronic ground state given the nuclear positions are treated in a unified way through an extended Lagrangian. The central idea in this approach is to define equations of motion for both the nuclei and the electronic degrees of freedom that are *simultaneously* followed using molecular dynamics. Car and Parrinello's extended Lagrangian is cleverly constructed with nuclear equations of motion similar to Eq. (9.16) and the introduction of the electronic degrees of freedom as fictitious dynamical variables. Schematically, this extended Lagrangian is

$$L = \frac{1}{2}\sum_{i=1}^{3N} m_i v_i^2 - E[\varphi(r_1,\ldots,r_{3N})] + \frac{1}{2}\sum_j 2\mu \int d\mathbf{r} |\dot{\psi}_j(\mathbf{r})|^2 + L_{\text{ortho}}. \quad (9.17)$$

The first two terms on the right-hand side are the same as in Eq. (9.16), while the last two terms introduce fictitious degrees of freedom. The third term that has the form of kinetic energy introduces a fictitious mass, μ, while the final term above is required to keep the one-electron wave functions orthogonal. When the velocities associated with the dynamics based on Eq. (9.17) are used to assign a temperature and scaled to bring $T \to 0$, the equilibrium state of minimal E is reached and the Lagrangian describes a real physical system on the potential energy surface. This method is referred to as Car–Parrinello molecular dynamics (CPMD).

Because the nuclear and electronic degrees of freedom are propagated simultaneously during a CPMD calculation, the total energy that is calculated at each time step does not correspond exactly to the true Born–Oppenheimer potential energy surface for the nuclear coordinates. This idea is illustrated in Fig. 9.1, which schematically shows the instantaneous electronic energy observed during a CPMD simulation. It is also important to realize that the dynamics of the electronic degrees of freedom during CPMD cannot be interpreted physically as the dynamics of electrons; the equations of motion for the electrons are merely a mathematical device to allow the dynamics of the nuclei to be generated in a numerically efficient way.

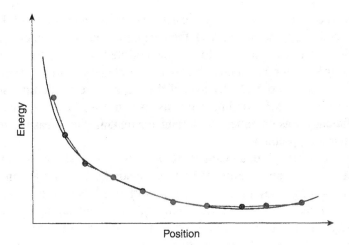

Figure 9.1 Schematic illustration of a CPMD trajectory along a single spatial coordinate. The smooth curve denotes the exact electronic ground-state energy, while the symbols indicate the electronic energy calculated by the CPMD method at each time step of an MD simulation.

At the time it was introduced, the Car–Parrinello method was adopted for *ab initio* molecular dynamics as well as for determination of the Kohn–Sham ground state (because their technique was faster than contemporary methods for matrix diagonalization). This situation changed in the 1990s, as a variety of efficient numerical methods for solving the Kohn–Sham equations based on iterative linear algebra methods were developed and widely applied. When used appropriately, the electronic information from a previous MD step can provide a good initial approximation for the ground state of the updated nuclear positions, enabling the energy and forces for a new time step to be computed in an efficient way. MD methods based on this approach are often referred to as Born–Oppenheimer molecular dynamics (BOMD) because they directly explore dynamics of nuclei on the Born–Oppenheimer potential energy surface. Although the extended Lagrangian approach of Car and Parrinello remains influential, calculations based on direct minimization of the Kohn–Sham equations at each time step are now more widely used.

To conclude our brief overview of *ab initio* MD, we note that the dynamics defined by Eq. (9.16) define a microcanonical ensemble. That is, trajectories defined by this Lagrangian will conserve the total energy of the system. Similar to the situation for classical MD simulations, it is often more useful to calculate trajectories associated with dynamics at a constant temperature. One common and effective way to do this is to add additional terms to the Lagrangian so that calculations can be done in the canonical ensemble (constant N, V, and T) using the Nosé–Hoover thermostat introduced in Section 9.1.2.

9.3 APPLICATIONS OF *AB INITIO* MOLECULAR DYNAMICS

9.3.1 Exploring Structurally Complex Materials: Liquids and Amorphous Phases

If you think back through all the preceding chapters of this book, you might notice that almost every calculation of a bulk material we have discussed has dealt with crystalline materials. One of the pragmatic reasons for this is that defining the positions of atoms in a crystalline material is straightforward. There are clearly many situations, however, where noncrystalline states of matter are interesting, including liquids and amorphous materials. Using DFT to examine these disordered states is challenging because any treatment of their material properties with DFT must be preceded by finding the organization of atoms in the material. *Ab initio* MD offers a powerful tool to tackle this problem.

We will illustrate the application of ab initio MD to liquid and amorphous phases using results from a detailed study by Lewis, De Vita, and Car of indium phosphide (InP).[1] This work was motivated by interest in using amorphous InP in the fabrication of microelectronic circuits. The idea for computationally generating models of liquid and amorphous materials closely mimics the approach that would be used for the same problem experimentally: a sample is heated to very high temperatures (to make a liquid); then the temperature is rapidly quenched (to form an amorphous material). Lewis et al. did calculations using a 64-atom supercell (32 In and 32 P atoms), with k space sampled at the Γ point. Molecular dynamics simulations were performed using a time step of 0.25 fs. The calculations began with the equilibrium crystal structure (the zinc-blende structure). The temperature history of the calculations is shown in Fig. 9.2. Note that this figure corresponds to 360,000 MD steps. Although this corresponds to a very time-consuming calculation, the quenching rate in the simulation, $\sim 10^{13}$ K/s is very large compared to the quenching rates typical in real processing of amorphous materials. In this calculation, the supercell dimensions were adjusted to give a liquid density 1.07 times larger than the crystal density and an amorphous density equal to the crystal density.

The energy of the InP system as a function of temperature during the quenching portion of the simulation is shown in Fig. 9.3. The change in slope of the energy as a function of temperature that can be seen at around 900 K is a characteristic of a liquid–glass transition. The final structure of the amorphous phase is a local minimum in energy because the temperature was quenched to 0 K, but this structure has ~ 0.24 eV/atom more energy than the crystalline material.

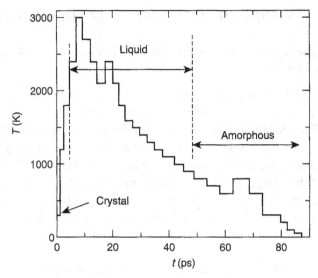

Figure 9.2 Temperature history of a CPMD simulation of InP indicating the melting of the crystal into a liquid and the quenching of the liquid into an amorphous solid. [Reprinted by permission from L. J. Lewis, A. De Vita, and R. Car, Structure and Electronic Properties of Amorphous Indium Phosphide from First Principles, *Phys. Rev. B* **57** (1998), 1594 (Copyright 1994 by the American Physical Society).]

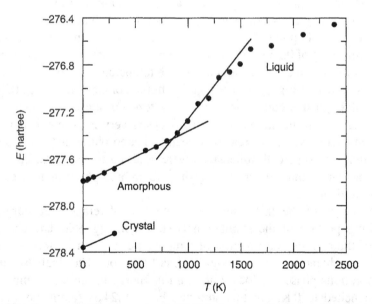

Figure 9.3 Energy of a 64-atom supercell of InP during quenching from a liquid state into an amorphous state. The energy of crystalline InP is also shown. (Reprinted by permission from the source cited in Fig. 9.2.)

Once the structures of liquid and amorphous InP have been generated using MD simulations, various properties of these materials can be explored. To give just one example, the structure of disordered materials is often characterized using the radial distribution function:

$$g(r) = \frac{\rho(r)}{4\pi r^2 \bar{\rho}}. \tag{9.18}$$

Here, $\rho(r)$ is the average density of atoms found in a thin shell at a radius r from an arbitrary atom in the material, and $\bar{\rho}$ is the average density of the entire material. For very small values of r, $g(r) \rightarrow 0$, since atoms cannot overlap one another. For large values of r, on the other hand, $g(r) \rightarrow 1$, because atoms that are separated from one another by large distances in a disordered material are not influenced by one another. The distribution functions calculated by Lewis et al. for liquid and amorphous InP are shown in Fig. 9.4. As might be expected, the amorphous material has considerably more structure than the liquid. One important feature in the amorphous material is the peak in the P–P distribution near $r = 2.2$ Å. This peak shows the existence of P–P bonds in the amorphous material, a kind of bond that does not exist in the crystalline solid.

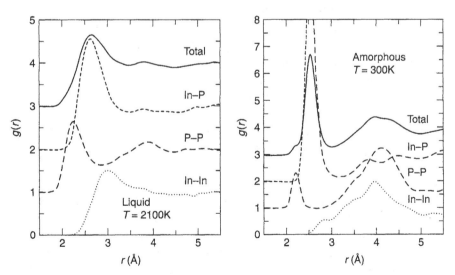

Figure 9.4   Radial distribution functions for liquid (left) and amorphous (right) InP. In each case, distribution functions for In–In, P–P, In–P atom pairs and for all atoms are shown. The zeros for the P–P, In–P, and total distribution functions are displaced vertically by 1, 2, and 3 units, respectively. (Reprinted by permission from the source cited in Fig. 9.2.)

9.3.2 Exploring Complex Energy Surfaces

In Chapter 2, our first example of a DFT calculation was to understand the crystal structure of a simple metal by comparing the energy of several candidate crystal structures. We will now consider the related problem of predicting the structure of a metal nanoparticle. Metal nanoparticles are extremely important in heterogeneous catalysis, and in recent years they have been used in a wide range of other technological applications, so there are strong motivations to understand their properties. A reasonable starting point for studying nanoparticles with DFT (or any other theoretical method) is to determine a particle's atomic structure. An important characteristic of this problem is that large numbers of different local minima can exist with quite different structures. The DFT calculations you are now familiar with are well suited to optimizing these structures once an initial approximation for the geometry of the nanoparticle is known, but calculations like this are only useful if a variety of initial structures can be generated.

One way to generate a variety of candidate nanoparticle structures is to use physical intuition or a search through the relevant literature to come up with plausible structures. This approach is equivalent to our approach for bulk materials of examining several well-known crystal structures. A severe limitation of this approach is that the number of atomic arrangements that can be contemplated grows very rapidly as the number of atoms is increased. Another limitation is that calculations performed in this way are inherently biased toward "familiar" structures. If the most stable nanoparticle has an unexpected form, then this may be difficult to find. These properties are characteristic of a range of physical problems that are defined by complex energy surfaces with large numbers of local energy minima.

Ab initio MD offers a useful method to complement searches of complex energy surfaces based on "rounding up the usual suspects." In this application of MD, the aim is not necessarily to generate a trajectory that faithfully remains on the Born–Oppenheimer potential energy surface. Instead, large time steps can be used in conjunction with a high-temperature simulation with the aim of rapidly generating an approximate trajectory that is able to overcome energy barriers between local energy minima.

Before embarking on any calculations of this type, it is useful to consider a rough estimate of how long we might expect to wait for events of interest to occur. If we are hoping to see an event that can be described as hopping over an energy barrier, then we can approximate the rate of this process using transition state theory as $k = \nu \exp(-\Delta E / k_B T)$, and the average time it will take to observe this event is $\tau = 1/k$. If we are interested in hopping between minima separated by energy barriers of ~ 1 eV and we estimate ν using the standard value of 10^{13} s^{-1}, then $\tau \sim 10^{-8}$ s at 1000 K and 10^{-11} s

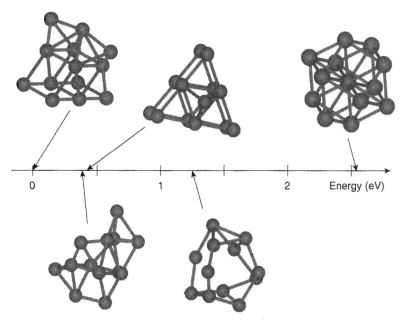

Figure 9.5 Five configurations of a Pt_{13} cluster that are local minima on this cluster's potential energy surface as characterized by GGA DFT calculations. The three clusters in the upper part of the figure were generated "by hand" based on symmetry considerations. The two clusters in the lower part of the figure were obtained using ab initio MD as described in the text. The energy of each cluster is defined relative to the lowest energy cluster.

at 2000 K. This means that an MD simulation (in which time steps are measured in femtoseconds) would be completely fruitless if performed at 1000 K but may yield a useful result at 2000 K.

As an example of using *ab initio* MD to explore a potential energy surface, we used this approach to find structures of a nanocluster of 13 Pt atoms.[†] Figure 9.5 shows three Pt_{13} nanoclusters with high levels of symmetry that were constructed "by hand" and then optimized with DFT. The cluster that is roughly spherical has O_h symmetry and is essentially a 13-atom fragment of the bulk fcc crystal structure. Intuitively, it may seem that this cluster would be the preferred structure for 13 Pt atoms, but the calculations indicate that this is not so. In the calculations shown in Fig. 9.5, the cluster with O_h symmetry is more than 2.5 eV higher in energy than the lowest energy configuration.

[†]This example is directly based on the extensive treatment of transition-metal nanoclusters in L.-L. Wang and D. D. Johnson, Density Functional Study of Structural Trends for Late-Transition-Metal 13-Atom Clusters, *Phys. Rev. B* **75** (2007), 235405.

To search for additional configurations of Pt_{13} that are energy minima, we followed an *ab initio* MD trajectory for 3 ps at 2000 K using time steps of 20 fs with the symmetric cluster with O_h symmetry as the initial condition. In this calculation, a Nosé–Hoover (NH) thermostat was used to control the temperature. The energy of the nanocluster (not including kinetic energy) after each MD step in this calculation is shown in Fig. 9.6. The figure shows that the nanocluster samples configurations with a large range of energies during this MD trajectory. To find local minima on the potential energy surface for Pt_{13}, a number of lower energy configurations observed during MD were optimized using standard energy minimization methods. When this was done for the states observed after steps 47, 95, and 147 in the trajectory shown in Fig. 9.6, two distinct local minima were found. These two minima are illustrated in Fig. 9.5. Both of these structures are quite asymmetric; so it would have been essentially impossible to generate them "by hand." More importantly, they have energies that are similar to the lowest energy symmetric clusters. In Wang and Johnson's comprehensive study of this system, they identified 18 distinct local minima with energies in the range shown in Fig. 9.5. Twelve of these structures were generated based on symmetric structures and the remainder were found using *ab initio* MD.

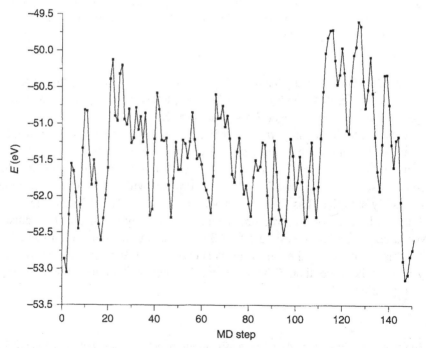

Figure 9.6 Energy of the Pt_{13} nanocluster simulated with *ab initio* MD using the methods described in the text.

Although using *ab initio* MD in the way we have just described can generate local minima that may be difficult to find in other ways, no guarantees can be given in calculations of this kind that all physically relevant minima have been found. This caveat also applies, of course, to searches for energy minima based on symmetry considerations or catalogs of anticipated structures. As long as the structures identified through MD simulations are interpreted while keeping this restriction in mind, *ab initio* MD can be a powerful tool for the exploration of complicated potential energy surfaces.

EXERCISES

1. In one of the exercises in Chapter 6, you estimated the activation energy for hopping of a charge-neutral vacancy in bulk Si. Use the same supercell to perform *ab initio* MD for this system, sampling reciprocal space only at the Γ point. First, determine what range of MD time steps can be used to generate trajectories that accurately conserve the system's total energy (electronic plus kinetic) using constant energy MD. Then, perform a constant temperature MD simulation at 1000 K. Observe whether the vacancy hops during your simulation and consider whether the outcome is consistent with your expectations based on your earlier estimate of the activation energy for this process.

2. Calculate the energy of the lowest and highest energy structures for Pt_{13} shown in Fig. 9.5. Also calculate the energy for at least one more of the symmetric structures identified by Wang and Johnson for this nanocluster. Use ab initio MD with the lowest energy symmetric structure as the starting point to find at least one nonsymmetric local minimum for Pt_{13}.

REFERENCE

1. L. J. Lewis, A. De Vita, and R. Car, Structure and Electronic Properties of Amorphous Indium Phosphide from First Principles, *Phys. Rev. B* **57** (1998), 1594.

FURTHER READING

Two excellent sources for learning about MD algorithms in classical simulations are M. P. Allen and D. J. Tildesley, *Computer Simulation of Liquids*, Clarendon, Oxford, UK, 1987, and D. Frenkel and B. Smit, *Understanding Molecular Simulations: From Algorithms to Applications*, 2nd ed., Academic, San Diego, 2002.

For further details on the algorithms underlying *ab initio* MD, see M. C. Payne, M. P. Teter, D. C. Allan, T. A. Arias, and J. D. Joannopoulos, *Rev. Mod. Phys.* **64** (1992), 1045, and R. M. Martin, *Electronic Structure: Basic Theory and Practical Methods*, Cambridge University Press, Cambridge, 2004.

The development of improved *ab initio* MD algorithms using DFT remains an active area. For one example of work in this area, see T. D. Kühne, M. Krack, F. R. Mohamed, and M. Parrinello, *Phys. Rev. Lett.* **98** (2007), 066401. Similar work exists for performing MD using high-level quantum chemistry techniques, as described, for example, in J. M. Herbert and M. Head-Gordon, *Phys. Chem. Chem. Phys.* **7** (2005), 3629.

The following studies give several applications of ab initio MD that sample the very large literature on this topic. For a discussion of simulating supercooled liquid alloys, see M. Widom, P. Ganesh, S. Kazimirov, D. Louca, and M. Mihalkovič, *J. Phys.: Condens. Matter*, **20** (2008), 114114. An example of comparing NEB and MD calculations to characterize atomic migration in a solid is given in Z. C. Wang, T. K. Gu, T. Kadohira, T. Tada, and S. Watanabe, *J. Chem. Phys.* **128** (2008), 014704. A study of liquid–liquid phase transformations in Si is described in N. Jakse and A. Pasturel, *Phys. Rev. Lett.* **99** (2007), 205702. An example of using configurations generated with MD to compute ensemble-averaged properties, specifically, the optical response of bulk GaAs, is given in Z. A. Ibrahim, A. I. Shkrebtii, M. J. G. Lee, K. Vynck, T. Teatro, W. Richter, T. Trepk, and T. Zettler, *Phys. Rev. B* **77** (2008), 125218. The accuracy of MD calculations using a GGA functional for reproducing the radial distribution functions of a variety of liquid metals is discussed in G. Kresse, *J. Non-crystalline Solids* **312–314** (2002), 52.

APPENDIX   CALCULATION DETAILS

All calculations for Pt_{13} clusters used the PW91 GGA functional and a cutoff energy of 191 eV. Each calculation placed a single Pt_{13} cluster in a cubic supercell with side length of 20 Å. Reciprocal space was sampled only at the Γ point, and the Methfessel–Paxton scheme was used with a smearing width of 0.1 eV. During geometry optimization, all atoms were allowed to relax until the forces on every atom were less than 0.03 eV/Å.

10

ACCURACY AND METHODS BEYOND "STANDARD" CALCULATIONS

10.1 HOW ACCURATE ARE DFT CALCULATIONS?

Throughout this book, we have focused on *using* plane-wave DFT calculations. As soon as you start performing these calculations, you should become curious about how accurate these calculations are. After all, the aim of doing electronic structure calculations is to predict material properties that can be compared with experimental measurements that may or may not have already been performed. Any time you present results from DFT calculations in any kind of scientific venue, you should expect someone in the audience to ask "How accurate are DFT calculations?"

So how accurate are DFT calculations? It is extremely important to recognize that despite the apparent simplicity of this question, it is not well posed. The notion of accuracy includes multiple ideas that need to be considered separately. In particular, it is useful to distinguish between *physical accuracy* and *numerical accuracy*. When discussing physical accuracy, we aim to understand how precise the predictions of a DFT calculation for a specific physical property are relative to the true value of that property as it would be measured in a (often hypothetical) perfect experimental measurement. In contrast, numerical accuracy assesses whether a calculation provides a well-converged numerical solution to the mathematical problem defined by the Kohn–Sham (KS) equations. If you perform DFT calculations, much of your day-to-day

Density Functional Theory: A Practical Introduction. By David S. Sholl and Janice A. Steckel
Copyright © 2009 John Wiley & Sons, Inc.

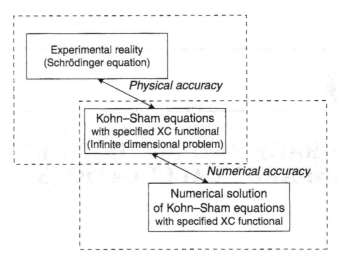

Figure 10.1   Illustration of the distinction between physical and numerical accuracy.

attention needs to be focused on issues of numerical accuracy. In making connections between your calculations and larger scientific conclusions, however, the various aspects of physical accuracy are of paramount importance. These two facets of accuracy are shown schematically in Fig. 10.1.

First, we will discuss the issue of numerical accuracy. The heart of any DFT calculation is to calculate the ground-state energy for the electrons associated with a set of atomic nuclei that exist at well-defined positions. Mathematically, our aim in a plane-wave calculation is to solve the KS equations, Eq. (1.5), for a spatially periodic collection of atomic nuclei using a specific exchange–correlation functional. Formally, this is an infinite-dimensional problem because the single-electron wave functions in the KS equations are continuous functions of the spatial coordinates. Because we must find solutions on a computer using a finite amount of resources in a finite amount of time, we must seek an approximate numerical solution.* To assess the numerical accuracy of a calculation, it is best to separately examine the approximations that are invoked to make the calculation computationally tractable. For plane-wave DFT calculations, these include the number and density of k points used to sample k space and the energy cutoff used to define the size of the plane-wave basis set. The implications of both of these choices are easy to understand—a large enough number of k points and a high enough energy cutoff give numerical solutions that converge to the true mathematical solution. Even when the number of k points and the energy cutoff are fixed, the KS equations must still be solved iteratively, so the outcome of this iterative

---

*This state of affairs is, of course, not unique to DFT calculations; it exists in any area where solutions to integral or differential equations that cannot be solved analytically are needed.

procedure converges toward a self-consistent solution of the equations only until a specified convergence criterion is satisfied. These issues were discussed at length in Chapter 3. It is important to note that demanding higher levels of numerical accuracy inevitably comes at the price of increased computational time. In many problems of practical interest a sensible trade-off between these competing demands must be selected.

A slightly more subtle aspect of numerical accuracy is important if we generalize the mathematical problem to finding the minimum ground-state energy of the electrons *and* nuclei as defined by the KS equations and a chosen exchange–correlation functional. To solve this problem, a numerical solution to the KS equations is found for the current positions of the nuclei, and an iterative optimization method is used to update the positions of the nuclei. The general properties of this important numerical problem were summarized in Section 3.3.1. Perhaps the most important feature of these calculations is that they can give reliable results for finding *local* minima in the total energy as the positions of the nuclei vary; they do not provide any guarantee that the *global* energy minimum has been found.

In many cases, a secondary quantity that can be defined in terms of the ground-state energy is of more interest than the energy itself. Issues of numerical accuracy also exist for many of these quantities. To give just one specific example, look back at the calculation of vibrational frequencies in Chapter 5 (in particular, Fig. 5.1). To numerically determine vibrational frequencies, a finite-difference approximation to a derivative of the total energy is used. At a formal level, this finite-difference approximation becomes exact in the limit that the step size used for the approximation vanishes. This implies that step sizes that are too large will give inaccurate results. But if step sizes that are too small are used, numerical inaccuracy appears from another source, namely the finite precision associated with the iterative solution of the KS equations. This is an example of the more general idea that it is important to understand the sources of numerical imprecision in your calculations in order to control the numerical accuracy of your results.

This discussion of numerical accuracy leads to several principles that should be applied when reporting DFT results:

1. *Accurately define the mathematical problem that was solved by specifying the exchange–correlation functional that was used.* Many different functionals exist; so it is not sufficient to state the kind of functional (LDA, GGA, etc.)—the functional must be identified precisely.
2. *Specify the numerical details used in performing the calculations.* A nonexhaustive list of these details include the size and shape of the supercell, the number and location of the k points, the method and

convergence criterion for iteratively solving the KS equations, and the method and convergence criterion for optimization of the nuclear positions. A useful metric for listing these properties is that access to this information should make it straightforward for someone on the other side of the globe to reproduce your calculations and get precisely the same numerical results.

3. *If you need to compare results from calculations with different levels of numerical approximations, do so with great caution.* For example, calculating energy differences between calculations using different numbers of k points introduces a systematic error into the result. Whenever this situation can be avoided by performing all calculations with the same numerical methods, you should do so. When this situation is unavoidable, as it may be when comparing your results to published results based on numerical approaches that are unavailable or impractical for you, clearly describe the differences between calculations that are being compared.

4. *When unambiguous comparisons between your results and experimental observations can be made, they should be.*

The way in which you address these principles in reporting your work should vary depending on the venue. In a technical paper, there is rarely any reason not to provide a detailed discussion of principles 1 and 2, at least in the supplementary information that almost all journals make freely available. On behalf of conference and seminar attendees everywhere, however, we beg you not to address principle number 2 from our list when you give a talk by reading through a long list of numerical specifications that are listed on a slide in a font requiring binoculars to be viewed legibly from the back of the room. Unless you are addressing a very unusual audience, a great majority of people in the room are mainly interested in the physical implications of your results and only want to be convinced that you are careful enough to generate reliable information! A good strategy for this situation is to prepare a slide with all the numerical details that you either show very briefly or reserve as an extra slide to use when answering questions on this topic.

We now turn to the issue of physical accuracy. Just as "accuracy" should not be considered as a single topic, "physical accuracy" is too vague of an idea to be coherently considered. It is much better to ask how accurate the predictions of DFT are for a specific property of interest. This approach recognizes that DFT results may be quite accurate for some physical properties but relatively inaccurate for other physical properties of the same material. To give just one example, plane-wave DFT calculations accurately predict the geometry of crystalline silicon and correctly classify this material as a semiconductor, but as we saw in Chapter 8 they overestimate the band gap of silicon by a considerable amount.

A second important aspect of physical accuracy has a decidedly sociological tinge to it: How accurately does the property of interest need to be predicted in order to make a useful scientific judgment? The answer to this important question varies significantly across scientific disciplines and even subdisciplines and depends strongly on what the information from a calculation will be used for. The answer also evolves over time; studies that were the state of the art 5, 10, or 20 years ago often become superceded by new approaches. This evolution of ideas due to the appearance of new methods is probably more rapid in computational modeling of materials than in experimental arenas, although it is important in both areas.

In many applied areas, calculations can be used to screen a variety of materials to select materials with useful properties. In this case, if DFT calculations (or any other kind of theoretical method) can reliably predict that material A is significantly better than material B, C, D, or E for some numerical figure of merit, then they can be extremely useful. That is, the accuracy with which trends in a property among various materials can be predicted may be more important than the absolute numerical values of the property for a specific material. In other venues, however, the precision of a prediction for a specific property of a single material may be more important. If calculations are being used to aid in assigning the modes observed in the vibrational spectrum of a complex material, for example, it is important to understand whether the calculated results should be expected to lie within 0.01, 1, or 100 cm^{-1} of the true modes.

To make meaningful judgments about this second aspect of physical accuracy, you must become familiar with the state-of-the-art literature (both computational and experimental) in your field of interest. In doing so, do not be afraid to make value judgments about how "good" the information you find is. If you are not able to make some decisions about which studies in an area are more significant than others, you will probably also have a difficult time making good decisions about how to proceed with your own calculations.[†]

In considering the application of DFT calculations to any area of applied science or engineering, it is important to honestly assess how relevant the physical properties accessible in these calculations are to the applications of interest. One useful way to do this is to list the physical phenomena that may influence the performance of a "real" material that are not included in a DFT calculation. We can assure you that it is far better to thoughtfully go through this process

---

[†]As one of the fictional rock musicians in the movie *Spinal Tap* said, "I believe virtually everything I read, and I think that is what makes me more of a selective human than someone who doesn't believe anything."

yourself than to have it done for you publicly by a skeptical experimenter at the conclusion of one of your talks or requests for funding!

As an example, imagine that we aim to use DFT calculations to advance the practice of heterogeneous catalysis for some large-volume chemical. There is a large scientific community devoting its resources to this task. A "standard" DFT calculation would examine the adsorption and reactions of key chemical intermediates on an extended surface of the metal being considered as a catalyst surrounded by a gas-phase region approximated by a vacuum. Many careful comparisons between highly controlled experiments and calculations have established that DFT can provide many powerful, quantitative insights into this situation. But an industrial heterogeneous catalyst is typically a metallic (or bimetallic) nanoparticle on an oxide support with complex local and microscale structure. The catalyst is often doped with "promoters" such as alkali atoms that have been found empirically to improve performance. Furthermore, the catalyst is operated at moderate to high pressures in either a complex mixture of gases or in a solvent, and the desired reaction is typically just one reaction among a large network of possible reactions generating less desirable by-products. By making judicious choices, it is possible to use calculations to examine some of the complexities of these real materials, but not all of them, and certainly not all of them simultaneously. This does not mean that calculations have no value in this area,[‡] but illustrates that understanding the context in which calculations can be used is an important skill to develop.

The preceding discussion highlights a tension that almost always exists when considering the physical accuracy of DFT calculations and quantum chemistry calculations in general. On one hand, it is typically desirable for a narrowly defined problem to move from less accurate calculation methods to more accurate methods, with the aim of predicting the "true" result. The almost inevitable result of this progression is a rapid increase in the computational effort required as the level of the theory is increased. On the other hand, broader scientific concerns often mean that there is a compelling need to look at systems defined by larger and more complex collections of atoms. Increasing system size also brings a heavy burden in terms of additional computational resources. The most useful scientific contributions based on calculations come from efforts that carefully balance the tension between these demands by understanding the state of the art of current knowledge in the area, the physical phenomena that are germane to the problem being studied, and the capabilities and limitations of the particular calculation methods that are used.

[‡] Karl Popper, the famous philosopher of science, said "Science may be described as the art of systematic oversimplification."

10.2 CHOOSING A FUNCTIONAL

The diagram in Fig. 10.1 emphasizes that density functional theory only describes a precisely defined mathematical problem once the exchange–correlation functional has been specified. Before choosing a functional to use, it is crucial that you understand the similarities and differences between the various functionals that are in common use. One useful classification of functionals has been described by John Perdew and co-workers[1] with reference to the Biblical account of Jacob in Genesis 28, where Jacob "... had a dream ... a ladder was set on the earth with its top stretching to heaven ... and the angels of God were ascending and descending on it." Jacob's ladder, as it illustrates this categorization of functionals for DFT, is illustrated in Fig. 10.2. Higher rungs represent functionals that include more and more physical information as steps toward the "perfect method" in which the Schrödinger equation is solved without approximation. To complete the

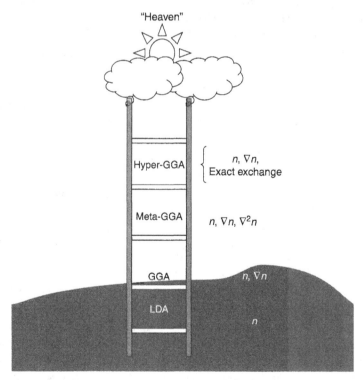

Figure 10.2   Illustration of Perdew's classification of DFT functionals using "Jacob's ladder," reaching rung by rung between Earth and "Heaven." The physical ingredients included in the functionals from each rung are summarized on the right.

metaphor, people applying DFT calculations are the angels moving back and forth between functionals as appropriate.

Although Jacob's ladder is a useful way to categorize functionals, other considerations are also important in making distinctions between the many functionals that are in common use. Perhaps the most important of these is whether a given functional is nonempirical or empirical. Nonempirical functionals have been constructed to satisfy certain constraints in order to make use of the known exact constraints on the true Kohn–Sham functional. Nonempirical functionals from higher rungs on the ladder in Fig. 10.2 have more ingredients and concurrently more constraints to satisfy. For nonempirical functionals, it is reasonable to expect (generally speaking) that accuracy will increase as additional constraints are satisfied. Empirical functionals, in contrast, are fitted to selected experimental or *ab initio* data and, therefore, contain a number of parameters that have been introduced and adjusted during the fitting process. Empirical functionals may (and often do) violate known properties of the true Kohn–Sham functional. This does not imply that semiempirical functionals are not accurate; they can produce very accurate results for certain problems, particularly those problems that are closely related to the systems included in the training set. However, it is not always reasonable to expect that going from a lower rung of the ladder in Fig. 10.2 to a higher rung will necessarily give better results, especially if the functional from the higher rung is an empirical functional.

In the remainder of this section, we give a brief overview of some of the functionals that are most widely used in plane-wave DFT calculations by examining each of the different approaches identified in Fig. 10.2 in turn. The simplest approximation to the true Kohn–Sham functional is the *local density approximation* (LDA). In the LDA, the local exchange–correlation potential in the Kohn–Sham equations [Eq. (1.5)] is defined as the exchange potential for the spatially uniform electron gas with the same density as the local electron density:

$$V_{XC}^{LDA}(\mathbf{r}) = V_{XC}^{electron\ gas}[n(\mathbf{r})]. \tag{10.1}$$

The exchange–correlation functional for the uniform electron gas is known to high precision for all values of the electron density, n. For some regimes, these results were determined from careful quantum Monte Carlo calculations, a computationally intensive technique that can converge to the exact solution of the Schrödinger equation. Practical LDA functionals use a continuous function that accurately fits the known values of $V_{XC}^{electron\ gas}(n)$. Several different functionals of this type were developed when this approach was initially developed, but they are essentially equivalent because they all describe $V_{XC}^{electron\ gas}(n)$

accurately. The most fundamental constraint of the true Kohn–Sham functional is the uniform density limit, which simply says that in the limit of uniform density the exchange–correlation energy should be the exact energy of the uniform electron gas at that density. By construction, the LDA is exact in the uniform density limit. Because the valence electron density in many bulk materials is slowly varying, satisfaction of this constraint is a key to the success of any functional for bulk solids. The electron density in atoms and molecules is not, in general, slowly varying, and the LDA does not fare quite as well in predictions of the properties of atoms and molecules.

The next approximation to the Kohn–Sham functional is the *generalized gradient approximation* (GGA). The physical idea behind the GGA is simple; real electron densities are not uniform, so including information on the spatial variation in the electron density can create a functional with greater flexibility to describe real materials. The equations on which the GGA are based are valid for slowly varying densities. In the GGA, the exchange–correlation functional is expressed using both the local electron density and gradient in the electron density:

$$V_{XC}^{GGA}(\mathbf{r}) = V_{XC}[n(\mathbf{r}), \nabla n(\mathbf{r})]. \tag{10.2}$$

Nonempirical GGA functionals satisfy the uniform density limit. In addition, they satisfy several known, exact properties of the exchange–correlation hole. Two widely used nonempirical functionals that satisfy these properties are the Perdew–Wang 91 (PW91) functional and the Perdew–Burke–Ernzerhof (PBE) functional. Because GGA functionals include more physical ingredients than the LDA functional, it is often assumed that nonempirical GGA functionals should be more accurate than the LDA. This is quite often true, but there are exceptions. One example is in the calculation of the surface energy of transition metals and oxides.[2]

The third rung of Jacob's ladder is defined by meta-GGA functionals, which include information from $n(\mathbf{r})$, $\nabla n(\mathbf{r})$, and $\nabla^2 n(\mathbf{r})$. In practice, the kinetic energy density of the Kohn–Sham orbitals,

$$\tau(\mathbf{r}) = \frac{1}{2} \sum_{\text{occupied states}} |\nabla \varphi_i(\mathbf{r})|^2, \tag{10.3}$$

contains the same physical information as the Laplacian of the electron density, and using these quantities has a number of advantages; so $\tau(\mathbf{r})$ may be used in meta-GGA functionals instead of $\nabla^2 n(\mathbf{r})$. The Tao–Perdew–Staroverov–Scuseria (TPSS) functional is an example of a nonempirical meta-GGA functional.

The fourth rung of the ladder in Fig. 10.2 is important because the most common functionals used in quantum chemistry calculations with localized basis sets lie at this level. The exact exchange energy can be derived from the exchange energy density, which can be written in terms of the Kohn–Sham orbitals as

$$E^{\text{exchange}}(\mathbf{r}) = -\frac{1}{2n(\mathbf{r})} \int d^3 r' \frac{\left| \sum\limits_{\text{occupied states}} \varphi_i^*(\mathbf{r}')\varphi_i(\mathbf{r}) \right|^2}{|\mathbf{r} - \mathbf{r}'|}. \tag{10.4}$$

A critical feature of this quantity is that it is *nonlocal*, that is, a functional based on this quantity cannot be evaluated at one particular spatial location unless the electron density is known for *all* spatial locations. If you look back at the Kohn–Sham equations in Chapter 1, you can see that introducing this nonlocality into the exchange–correlation functional creates new numerical complications that are not present if a local functional is used. Functionals that include contributions from the exact exchange energy with a GGA functional are classified as hyper-GGAs.

Hyper-GGA functionals describe exchange using a mixture of the exact exchange and a GGA exchange functional. By far the most widely used of these functionals is the B3LYP functional, which can be written as

$$V_{\text{XC}}^{\text{B3LYP}} = V_{\text{XC}}^{\text{LDA}} + \alpha_1(E^{\text{exchange}} - V_X^{\text{LDA}}) + \alpha_2(V_X^{\text{GGA}} - V_X^{\text{LDA}})$$
$$+ \alpha_3(V_C^{\text{GGA}} - V_C^{\text{LDA}}). \tag{10.5}$$

Here, V_X^{GGA} is the Becke 88 exchange functional, V_C^{GGA} is the Lee–Yang–Parr correlation functional, and α_1, α_2, and α_3 are three numerical parameters. The three parameters were chosen empirically to optimize the performance of the functional for a sizable test set of molecular properties (bond lengths, formation energies, etc.). The B3LYP functional, and some other similar functionals, have been fantastically successful in predictions of properties of small molecules. It is important to note that B3LYP does not satisfy the uniform density limit. Thus, it would not be expected to perform especially well (as indeed it does not) in predictions for bulk materials, especially metals.

The classes of functionals shown in Fig. 10.2 do not exhaust the kinds of functionals that can be constructed. Any functional that incorporates exact exchange is called a hybrid functional. The hyper-GGA functionals listed above can therefore also be referred to as hybrid-GGA methods. Hybrid-meta-GGA functionals also exist—these combine exact exchange with meta-GGA functionals; this group of functionals is not shown in Fig. 10.2.

If Fig. 10.2 was taken too literally, the figure could be interpreted as implying that functionals from a higher rung are *always* better than functionals from a lower row. This is not necessarily the case! Ascending to a higher rung includes more physical ingredients in a functional, but each row involves systematic errors between DFT results and nature's true outcomes. These errors do not always become smaller for specific properties of interest in moving up the first few rows. Litimein et al., for example, examined several III–V nitrides with various nonempirical functionals.[3] The physical properties of the materials that they considered were best described by the TPSS functional, the nonempirical meta-GGA functional mentioned above, but LDA calculations typically had somewhat more accurate results than calculations with the PBE–GGA functional.

Even greater caution needs to be exercised in thinking about how Fig. 10.2 is applied to empirical functionals. By design, empirical functionals can work extremely well for physical situations that are chemically similar to the test set that was used in their development. Care needs to be taken, however, when applying empirical functionals to situations that are very different from these test sets. To give just one specific example, von Oertzen and co-workers compared LDA, PBE–GGA, and B3LYP-hyper–GGA calculations for the crystal and electronic structure for several mineral sulfides.[4] All three functionals predicted the lattice parameters of the sulfides within $1-2\%$ of the experimental values. For PbS, calculations of the band gap using LDA, PBE–GGA, and B3LYP gave errors of -0.1, $+0.1$, and $+0.7$ eV relative to experimentally observed values. For FeS_2, the two nonempirical functionals underpredicted the band gap by about 0.5 eV, while B3LYP gave a band gap that was 2.2 eV too large.

Hybrid functionals introduce a strong divide between DFT calculations based on localized basis sets and calculations based on plane waves. Because of the numerical details associated with solving the Kohn–Sham equations in a plane-wave basis set, introducing the nonlocality of exact exchange greatly increases the numerical burden of solving these equations. This difficulty is not so severe when localized basis sets are used. As a result, functionals making use of a portion of the exact exchange find almost universal use in the chemistry community within codes using local basis sets, while these functionals are currently very difficult to apply to calculations involving supercells with periodic boundary conditions and spatially extended materials. Exciting progress is being realized in the development of screened hybrid functionals. In this approach, the exchange interaction is split into two components, a long-range and a short-range one. A degree of the exact exchange is applied only to the short-range portion. The HSE functional is based on this approach; it is built by starting with the PBE functional and mixing a portion of exact exchange into only the short-range portion of the problem. The Further Reading section at the end of the chapter lists several resources for learning more about progress in this area.

To conclude this rapid tour of exchange–correlation functionals, we reiterate our comment above that you must spend time understanding the current state of the art of the literature in your area of interest in order to make judgments about what functionals are appropriate for your own work.

10.3   EXAMPLES OF PHYSICAL ACCURACY

10.3.1   Benchmark Calculations for Molecular Systems—Energy and Geometry

There is a large literature associated with the physical accuracy of DFT calculations (and other quantum chemistry methods) using localized basis sets for molecular quantum chemistry. Sousa, Fernandes, and Ramos recently reviewed this literature,[5] discussing dozens of benchmarking studies, many of which compared the performance of dozens of DFT functionals. Table 10.1 summarizes the typical results observed for calculations with the B3LYP functional, the most widely used functional for localized basis set calculations. Because this table condenses a huge number of calculations performed by many people into just a few lines, it is highly recommended that you consult the sources cited by Sousa, Fernandes, and Ramos[5] to understand in more detail the details associated with making these estimates.

A number of points need to be kept in mind when considering the values listed in Table 10.1. The precision of the results is characterized by the mean absolute error, that is, the mean of the absolute value of the difference

TABLE 10.1   Typical Accuracy for B3LYP DFT Calculations on Small Molecule Test Sets[a]

Physical Property	Typical Mean Absolute Error	Typical Value in Test Sets
Bond length	0.01 Å	1.1–1.5 Å
Bond angle	1°	110°
Barrier height	3–4 kcal/mol	12–25 kcal/mol
Atomization energies	1–2 kcal/mol (10–20 kcal/mol for metal dimers)	500 kcal/mol (58 kcal/mol for metal dimers)
Binding energies	5–10 kcal/mol	60–80 kcal/mol
Heats of formation	5–20 kcal/mol	50 kcal/mol
Hydrogen bond strengths	0.5–1 kcal/mol	8 kcal/mol

[a]Adapted from the literature review by Sousa, Fernandes, and Ramos.[5] Atomization energies, binding energies, and heats of formation are reported per molecule.

between calculated and experimental values. In essentially every test set there are examples where the error is significantly larger than the mean, and these cases are typically not randomly scattered through the test set. In other words, there are physical trends among molecules in the accuracy of B3LYP and other DFT calculations that can be obscured by focusing too closely on the mean absolute error. The atomization energies described in Table 10.1 are one simple example of this idea where the results for metal dimers are substantially less precise than the results for small molecules that do not include metals.

A second aspect of the results in Table 10.1 is that B3LYP does not necessarily give the best performance among all DFT functionals that have been tested. In the 55 benchmarking studies listed by Sousa, Fernandes, and Ramos,[5] B3LYP gave the best performance (as characterized by the mean absolute error) among the functionals tested in only four cases. In at least one case, B3LYP showed the worst performance among the group of functionals tested. There is no single functional, however, that emerges as the "best" for all or even most of the properties that have been examined in these benchmarking studies. The B3LYP results, however, are representative of the levels of accuracy that can be expected from carefully developed DFT functionals for localized basis set calculations.

10.3.2 Benchmark Calculations for Molecular Systems—Vibrational Frequencies

Harmonic vibrational frequencies computed by all types of quantum chemistry calculations (not just DFT) tend to systematically overestimate experimental data. This overestimation occurs in part because of anharmonicities in real vibrations, but also because of the inexact nature of calculated solutions to the Schrödinger equation. It is possible to calculate anharmonic corrections to vibrational frequencies by fitting the potential energy surface in the neighborhood of an energy minimum with a Taylor expansion that goes beyond the second-order (harmonic) terms, but this approach is only used rarely. A common approach in molecular quantum chemistry calculations is to note that the ratio between computed harmonic frequencies and experimentally observed frequencies tends to vary little over a large number of examples provided that the same level of theory is used in all calculations. This observation means that it is reasonable to interpret calculated vibrational frequencies by multiplying them by a scaling factor when making comparisons with experiments. Merrick and co-workers have tabulated an extensive collection of scaling factors using localized basis set calculations at many different levels of theory, including multiple GGA and hyper-GGA functionals.[6] These scaling factors are typically between 0.95 and 0.99; that is, vibrational frequencies are overestimated in these calculations by $1-5\%$. Although comparisons between vibrational frequencies computed using plane-wave DFT calculations and experimental data have not been made with the same level of detail as for localized basis set calculations, this range of accuracy also appears typical for plane-wave calculations.

10.3.3   Crystal Structures and Cohesive Energies

In the next two subsections, we describe collections of calculations that have been used to probe the physical accuracy of plane-wave DFT calculations. An important feature of plane-wave calculations is that they can be applied to bulk materials and other situations where the localized basis set approaches of molecular quantum chemistry are computationally impractical. To develop benchmarks for the performance of plane-wave methods for these properties, they must be compared with accurate experimental data. One of the reasons that benchmarking efforts for molecular quantum chemistry have been so successful is that very large collections of high-precision experimental data are available for small molecules. Data sets of similar size are not always available for the properties of interest in plane-wave DFT calculations, and this has limited the number of studies that have been performed with the aim of comparing predictions from plane-wave DFT with quantitative experimental information from a large number of materials. There are, of course, many hundreds of comparisons that have been made with individual experimental measurements. If you follow our advice and become familiar with the state-of-the-art literature in your particular area of interest, you will find examples of this kind. Below, we collect a number of examples where efforts have been made to compare the accuracy of plane-wave DFT calculations against systematic collections of experimental data.

Milman and co-workers optimized the structure of 148 inorganic crystals and molecular compounds that included 81 different elements from the periodic table using PW91 GGA calculations.[7] These calculations included oxides, nitrides, carbides, alkaline earth and transition metals, and semiconductors. In general, the lattice parameters for crystalline solids and bond lengths for molecular species had errors of 2% or less compared to experimental data, with both positive and negative deviations from experimental data. Only 5 of the 148 examples had errors larger than 3% relative to experiments. A similar level of accuracy has been reported by Alapati et al. in a series of PW91 GGA calculations for a large set of crystal structures relevant for studying light metal hydrides and borides.[8] The general observation from these studies is consistent with the observation from localized basis set calculations (see Table 10.1) that DFT calculations can reliably give accurate information about the geometry of a wide range of materials.

Philipsen and Baerends examined the cohesive energy and bulk modulus of 11 elemental solids from four columns of the periodic table using PW91 GGA calculations.[9] They found mean absolute errors of 0.35 eV and 0.15 Mbar for the cohesive energy and bulk modulus, respectively, for this set of materials. These errors are (on average) smaller than the analogous results for LDA calculations, although there are specific examples for which LDA results are in better agreement with experiment than GGA calculations. Korhonen and co-workers reported PW91 GGA calculations of the vacancy formation energy in 12 elemental metals.[10] For this property, there are differences of >0.3 eV between different experimental measurements for some, although certainly not all, metals. Using the mean of the reported experimental values as a basis for comparison, the calculated results show a mean absolute error of 0.30 eV for these 12 metals.

10.3.4 Adsorption Energies and Bond Strengths

The performance of various DFT functionals for predicting the adsorption energy of small molecules on metal surfaces has been systematically studied by a number of groups. Hammer, Hansen, and Nørskov examined CO adsorption on five metal surfaces and NO adsorption on four metal surfaces using five different functionals.[11] In general, DFT overpredicts these adsorption energies, with LDA calculations giving considerably worse results than the various GGA functionals. The PW91 GGA functional gave an RMS deviation between calculations and experimental data of 0.78 eV for CO and 0.52 eV for NO. This RMS deviation was reduced slightly to 0.67 and 0.43 eV, respectively, when the PBE GGA functional was used. Hammer et al. introduced a revision of the PBE functional now known as the RPBE GGA functional, which gave substantially better predictions for CO and NO adsorption energies than the earlier functionals, with RMS deviations of 0.37 and 0.22 eV, respectively.[12] The RPBE functional is generally considered to give an improved description of chemisorptions relative to the PW91 and PBE functionals.

A more extensive comparison of DFT-predicted adsorption energies with experimental data for CO adsorption on metal surfaces was made using data from 16 different metal surfaces by Abild-Pedersen and Andersson.[13] Unlike the earlier comparison by Hammer et al., this study included information on the uncertainties in experimentally measured adsorption energies by comparing multiple experiments for individual surfaces when possible. These uncertainties were estimated to be on the order of 0.1 eV for most surfaces. In situations where multiple experimental results were available, the mean of the reported experimental results was used for comparison with DFT results. For calculations with the PW91 GGA functional, the mean absolute deviation between the DFT and experimental adsorption energies was 0.48 eV, while the mean deviation was -0.46 eV, indicating that this functional overbinds CO in almost all cases. Similar calculations with the RPBE GGA functional gave a mean deviation of -0.12 eV and a mean absolute deviation of 0.25 eV. Abild-Pedersen and Andersson also tested a simple correction for the adsorption energy of CO that involves examining molecular orbital properties of CO as calculated with the DFT functional being used. Applying this correction to the RPBE GGA results gave a mean deviation in the overall CO adsorption energy of 0.06 eV. This is an encouraging sign that the systematic errors associated with using GGA functionals to predict molecular adsorption energies can be reduced, but this method is not yet well developed for the general problem of understanding molecular adsorption.

One general explanation for the systematic inaccuracies of DFT for adsorption energies is that these calculations inherently involve comparing chemically dissimilar states, namely an isolated molecule and a molecule chemically bound to a surface. An even simpler example of this situation is to consider the bond strength of simple molecules. As a specific example, Fuchs et al. (as part of a larger study) examined the bond energy of N_2 as predicted by various DFT functionals.[14] They found that the LDA functional overbinds N_2 by about 1 eV/atom. This overbinding is reduced to 0.5 and 0.2 eV/atom by using the PBE and RPBE GGA functionals, respectively. The imprecision in predicted bond energies is not a feature that appears only in plane-wave DFT calculations, as the information for molecular energies from localized basis set DFT calculations in Table 10.1 indicates.

The overprediction of binding energies for simple molecules can have consequences for *ab initio* thermodynamic calculations involving these species. In Chapter 7 we examined the stability of bulk oxide materials using this approach. In order to quantitatively predict the conditions at which bulk phases were in equilibrium, the chemical potential of O_2 had to be calculated using Eq. (7.6), which includes a contribution from the total energy of O_2. One way to correct for the imprecision of DFT for total energies of this kind is to use experimental data when it is available. When this approach is used, it is important that it is clearly stated.

10.4   DFT+X METHODS FOR IMPROVED TREATMENT OF ELECTRON CORRELATION

The discussion of the hierarchy of functionals in Section 10.2 highlighted the observation that while the exchange energy of a collection of electrons can, in principle, be evaluated exactly, the correlation energy cannot. As a result, it is reasonable to be particularly concerned about the reliability of DFT calculations for physical properties that are governed primarily by electron correlation. Two important examples of this kind come from what may seem like physically disparate situations; weak long-range interactions between atoms and the electronic structure of strongly correlated materials. Below, we introduce each of these topics and briefly describe "add-ons" to standard DFT that can be used to treat them.

10.4.1 Dispersion Interactions and DFT-D

Dispersion interactions (also known as van der Waals interactions or London forces) play an incredibly important role in our everyday lives. Consider the gasoline or diesel that all of us rely on as transportation fuels. The huge global infrastructure that exists to deliver and use these fuels relies on the fact that they are liquids. The fact that nonpolar molecules such as hexane readily form liquids is a signature of the net attractive interactions that exist between hexane molecules. These attractive interactions arise directly as a result of electron correlation.

The relationship between electron correlation and long-range forces between atoms was initially examined in the 1930s by London. He realized that although the time-average electron density around an atom or nonpolar molecule has no dipole moment, electron oscillations lead to deformations of the density resulting in a transient dipole moment. This instantaneous dipole moment can induce a temporary dipole moment on other atoms or molecules by distorting their electron density. The existence of two dipoles creates a net interaction. London showed that the general form of the interaction between two spherically symmetric atoms at large distances was

$$V^{\text{dispersion}} = -\frac{C}{r^6}, \tag{10.6}$$

where r is the distance between the atoms and C is a collection of physical constants.[§]

A simple physical example to illustrate dispersion interactions are the dimers of rare-gas atoms such as He, Ne, and Ar. These atoms are well known for their lack of chemical reactivity, but the fact that these gases can be liquefied at sufficiently low temperatures makes it clear that attractive interactions between rare-gas atoms exist. Zhao and Truhlar examined the performance of a large number of meta-GGA functionals for describing He–He, He–Ne, He–Ar, and Ne–Ar dimers with localized basis set calculations.[15]

[§]These physical constants include some of the same information that characterizes the dispersion of the refractive index of a material with respect to the wavelength of light; hence the description of these interactions as dispersion interactions.

The experimentally determined interaction energies of these dimers vary from 0.04 to 0.13 kcal/mol, and the equilibrium atom-to-atom distances vary from 3.03 to 3.49 Å. When the B3LYP functional was used, even the existence of a minimum in the potential well defined by the dimer depended on the details of the calculations. In calculations performed without counterpoise corrections, the equilibrium separations of the two atoms were predicted with a mean absolute error of 0.13 Å, while the mean absolute error in the interaction energy was 0.24 kcal/mol.

Another simple example illustrating the limitations of DFT calculations with respect to dispersion interactions is the crystal structures of molecular crystals. Neumann and Perrin examined the crystal structure of 20 molecular crystals using plane-wave DFT calculations with the PW91 GGA functional.[16] On average, the unit cell volume predicted with DFT was 20% larger than the experimental result. For example, the unit cell volume of crystalline butane (C_4H_{10}) was overestimated by 27%. The volumes of crystalline methanol (CH_3OH) and terephthalic acid $(C_6H_4(COOH)_2)$ were overestimated by 14% and 25%, respectively. These results are very different from the crystalline solids discussed above in Section 10.3.3, and the difficulties with the molecular crystals stem from the important contributions of dispersion interactions to the intermolecular forces in these materials.

One conceptually simple remedy for the shortcomings of DFT regarding dispersion forces is to simply add a dispersion-like contribution to the total energy between each pair of atoms in a material. This idea has been developed within localized basis set methods as the so-called DFT-D method. In DFT-D calculations, the total energy of a collection of atoms as calculated with DFT, E_{DFT}, is augmented as follows:

$$E_{DFT-D} = E_{DFT} - S \sum_{i \neq j} \frac{C_{ij}}{r_{ij}^6} f_{damp}(r_{ij}). \qquad (10.7)$$

Here, r_{ij} is the distance between atoms i and j, C_{ij} is a dispersion coefficient for atoms i and j, which can be calculated directly from tabulated properties of the individual atoms, and $f_{damp}(r_{ij})$ is a damping function to avoid unphysical behavior of the dispersion term for small distances. The only empirical parameter in this expression is S, a scaling factor that is applied uniformly to all pairs of atoms. In applications of DFT-D, this scaling factor has been estimated separately for each functional of interest by optimizing its value with respect to collections of molecular complexes in which dispersion interactions are important. There are no fundamental barriers to applying the ideas of DFT-D within plane-wave DFT calculations. In the work by Neumann and Perrin mentioned above, they showed that adding dispersion corrections to forces from plane-wave DFT calculations can yield crystal unit cell volumes for molecular crystals within roughly 1% of experimental results. These calculations were performed by coupling a plane-wave DFT code with a separate code designed for optimization of crystal structures. We are not aware of any plane-wave codes at present that include dispersion corrections as an "off the shelf" option.

10.4.2 Self-Interaction Error, Strongly Correlated Electron Systems, and DFT+U

Dispersion interactions are, roughly speaking, associated with interacting electrons that are well separated spatially. DFT also has a systematic difficulty that results from an unphysical interaction of an electron with itself. To understand the origin of the self-interaction error, it is useful to look at the Kohn–Sham equations. In the KS formulation, energy is calculated by solving a series of one-electron equations of the form

$$\left[-\frac{\hbar^2}{2m} \nabla^2 + V(\mathbf{r}) + V_H(\mathbf{r}) + V_{XC}(\mathbf{r}) \right] \psi_i(\mathbf{r}) = \varepsilon_i \psi_i(\mathbf{r}). \tag{10.8}$$

As described in Chapter 1, the first term on the left-hand side describes the kinetic energy of the electron, V is the potential energy of an electron interacting with the nuclei, V_H is the Hartree electron–electron repulsion potential, and V_{XC} is the exchange–correlation potential. This approach divides electron–electron interactions into a classical part, defined by the Hartree term, and everything else, which is lumped into the exchange–correlation term. The Hartree potential describes the Coulomb repulsion between the electron and the system's total electron density:

$$V_H(\mathbf{r}) = e^2 \int \frac{n(\mathbf{r}')}{|\mathbf{r} - \mathbf{r}'|} d^3 r'. \tag{10.9}$$

This potential includes an unphysical repulsive interaction between the electron and itself (because the electron contributes to the total electron density). The energy associated with this unphysical interaction is the self-interaction energy.

In the Hartree–Fock (HF) method, the spurious self-interaction energy in the Hartree potential is exactly cancelled by the contributions to the energy from exchange. This would also occur in DFT if we knew the exact Kohn–Sham functional. In any approximate DFT functional, however, a systematic error arises due to incomplete cancellation of the self-interaction energy.

The physical implications of the self-interaction error are easy to spot in H_2^+, a one-electron system. In this example there should be no electron–electron interactions at all, and yet the Hartree potential is nonzero. DFT calculations give a qualitatively incorrect description of the dissociation of H_2^+ because of the self-interaction error. In more general terms, self-interaction error causes Kohn–Sham orbitals that are highly localized to be improperly destabilized with approximate exchange–correlation functionals. Unpaired electrons tend to delocalize spatially in order to minimize self-interaction. When electronic states with many strongly localized electrons exist, these states are said to be strongly correlated. The best known materials of this kind are actinides and various transition-metal oxides that include partially filled d or f shells. Self-interaction errors can lead to DFT giving an incorrect $d^{n-1}s^1$ configuration instead of the $d^{n-2}s^2$ configuration in some $3d$ transition metals, which accounts for errors in the calculation of cohesive energies for certain materials.

Self-interaction error (SIE) can also play a role in the calculation of defects in solids. One example is the calculations by Pacchioni for an Al substitutional defect in SiO_2.[17] Spin density plots of the $[AlO_4]^0$ center that were generated using (SIE-free) HF calculations correctly show the hole trapped in a nonbonding $2p$ orbital of one O atom adjacent to the Al, in concurrence with experimental data. DFT calculations with some functionals incorrectly spread the hole over all four O atoms adjacent to the Al. This effect was reduced when hybrid functionals were used because the inclusion of exact exchange in these functionals partially corrects the self-interaction errors present in nonhybrid functionals. Another example of the effect of self-interaction error is that a magnetic insulator such as NiO is described as a metal with standard DFT functionals.[18] Hybrid exchange functionals partially correct the self-interaction error because of their (partial) use of the exact exchange energy. As discussed in Section 10.2, hybrid functionals are gradually becoming available in plane-wave codes as we write.

The fact that self-interaction errors are canceled exactly in HF calculations suggests that a judicious combination of an HF-like approach for localized states with DFT for "everything else" may be a viable approach for strongly correlated electron materials. This idea is the motivation for a group of methods known as DFT+U. The usual application of this method introduces a correction to the DFT energy that corrects for electron self-interaction by introducing a single numerical parameter, $U - J$, where U and J involve different aspects of self-interaction. The numerical tools needed to use DFT+U are now fairly widely implemented in plane-wave DFT codes.

The use of DFT+U obviously requires that the numerical value of the parameter $U - J$ be specified. There are two common approaches to this problem, both of which have merits and disadvantages. The first approach is to take a known property of some relevant material, typically a perfect crystal, and determine what value of $U - J$ gives the closest result to this property for a given functional. A second approach is to use some other kind of *ab initio* calculation in a test system where these calculations are feasible to estimate $U - J$. This approach tends to give bounds on $U - J$ rather than precise values. Because of the ambiguities associated with assigning the value of $U - J$, it is important when performing DFT+U calculations to understand the sensitivity of the results to the value of this parameter.

10.5   LARGER SYSTEM SIZES WITH LINEAR SCALING METHODS AND CLASSICAL FORCE FIELDS

It is not difficult to imagine examples where the ability to perform quantitative calculations involving thousands, tens of thousands, or even more atoms would be extremely useful. Your own experience with DFT calculations has by now taught you that reaching this goal with "standard" methods is very unlikely because the computational effort required for these calculations grows rapidly as the system size grows. To be more specific, the computational time needed to compute the total energy of a set of atoms[||] with the methods we have discussed throughout this book is

$$T = cN^3, \tag{10.10}$$

where N is the number of atoms and c is a prefactor that depends on various details of the method. This scaling is an asymptotic result that reflects the behavior of large systems, and it implies that an order of magnitude increase in computer speed only allows us to roughly double our system size if the computational time is fixed.

The scaling shown in Eq. (10.10) has motivated strenuous efforts to develop alternative calculation methods with better scaling. This work has focused on so-called linear scaling methods, for which the time needed for a single energy calculation scales as

$$T = c'N. \tag{10.11}$$

The advantage of methods obeying Eq. (10.11) over the scaling in Eq. (10.10) seem so obvious that these new methods would appear poised to sweep away the old once they are available. The situation is not quite this simple, however.

[||]The relevant time for a useful calculation involves both the time for a single energy calculation and also the number of energy calculations needed to address the physical question of interest.

A general property of linear scaling methods compared to the standard approaches is that when calculations with the same level of theory are compared, $c/c' < 1$. That is, the linear scaling approach is only faster than a standard calculation once the system size is large enough. With the linear scaling codes that are currently available for bulk systems, the crossover point at which the linear scaling approach become favorable occurs for systems containing hundreds of atoms, although this number varies depending on the system being studied. This means that linear scaling methods have not (yet) displaced standard calculations, although examples are starting to appear in which these methods are being used to tackle specific problems. It is possible, however, that linear scaling calculations will become more widely used in the future.

An alternative avenue for performing calculations involving very large numbers of atoms is to use methods based on classical force fields. Calculations of this kind involving hundreds of thousands or even millions of atoms are now relatively routine. When classical simulations are performed using MD, the overall time scales that are accessible are limited for the reasons discussed in Chapter 9, although these time scales are orders of magnitude longer than those accessible using *ab initio* MD. A large range of sophisticated Monte Carlo sampling techniques are available that allow sampling of a variety of physical properties that cannot be accessed using MD alone. DFT and other kinds of quantum chemistry calculations can play a crucial supporting role in advancing classical force field calculations by providing data to be used in developing the interatomic force fields that define these methods.

10.6   CONCLUSION

We hope that this book has given you an efficient introduction into the many uses of plane-wave DFT calculations in describing real materials. In any active area of science, it is dangerous to conclude that your understanding of any method is "complete," and this is certainly true for the topics covered in this book. We urge you to read widely and think critically as you perform DFT calculations or interpret calculations by your collaborators. We also urge you to learn about the other methods that together with plane-wave DFT form the core of current approaches to atomically detailed materials modeling, namely the quantum chemistry methods described briefly in Section 1.6 and the classical methods touched upon in the previous section. As you become adept at understanding the strengths and limitations of all of these approaches, you will be able to critically evaluate and positively contribute to the enormous range of scientific problems where detailed materials modeling now plays an inextricable role.

REFERENCES

1. J. P. Perdew, A. Ruzsinszky, J. M. Tao, V. N. Staroverov, G. E. Scuseria, and G. I. Csonka, Prescription for the Design and Selection of Density Functional Approximations: More Constraint Satisfaction with Fewer Fits, *J. Chem. Phys.* **123** (2005), 062201.

2. D. Alfé and M. J. Gillan, The Energetics of Oxide Surfaces by Quantum Monte Carlo, *J. Phys.: Condens. Matt.* **18** (2006), L435.

3. F. Litimein, B. Bouhafs, G. Nouet, and P. Ruterana, Meta-GGA Calculation of the Electronic Structure of Group III–V Nitrides, *Phys. Stat. Sol. B* **243** (2006), 1577.

4. G. U. von Oertzen, R. T. Jones, and A. R. Gerson, Electronic and Optical Properties of Fe, Zn and Pb Sulfides, *Phys. Chem. Minerals* **32** (2005), 255.

5. S. F. Sousa, P. A. Fernandes, and M. J. Ramos, General Performance of Density Functionals, *J. Phys. Chem. A* **111** (2007), 10439.

6. J. P. Merrick, D. Moran, and L. Radom, An Evaluation of Harmonic Vibrational Frequency Scale Factors, *J. Phys. Chem. A* **111** (2007), 11683.

7. V. Milman, B. Winkler, J. A. White, C. J. Pickard, M. C. Payne, E. V. Akhmatskaya, and R. H. Nobes, Electronic Structure, Properties, and Phase Stability of Inorganic Crystals: A Pseudopotential Plane-Wave Study, *Int. J. Quantum Chem.* **77** (2000), 895.

8. S. V. Alapati, J. K. Johnson, and D. S. Sholl, Using First Principles Calculations to Identify New Destabilized Metal Hydride Reactions for Reversible Hydrogen Storage, *Phys. Chem. Chem. Phys.* **9** (2007), 1438.

9. P. H. T. Philipsen and E. J. Baerends, Relativistic Calculations to Assess the Ability of the Generalized Gradient Approximation to Reproduce Trends in Cohesive Properties of Solids, *Phys. Rev. B* **61** (2000), 1773.

10. T. Korhonen, M. J. Puska, and R. M. Nieminen, Vacancy-Formation Energies for fcc and bcc Transition Metals, *Phys. Rev. B* **51** (1995), 9526.

11. B. Hammer, L. B. Hansen, and J. K. Nørskov, Improved Adsorption Energetic within Density-Functional Theory Using Revised Perdew–Burke–Ernzerhof Functionals, *Phys. Rev. B* **59** (1999), 7413.

12. Another revision to the PBE functional has been dubbed revPBE; see Y. K. Zhang and W. T. Yang, Comment on Gradient Approximation Made Simple, *Phys. Rev. Lett.* **80** (1998), 890, and the reply from the PBE authors: J. P. Perdew, K. Burke, and M. Ernzerhof, *Phys. Rev. Lett.* **80** (1998), 891.

13. F. Abild-Pedersen and M. P. Andersson, CO Adsorption Energies on Metals with Correction for High Coordination Adsorption Sites—A Density Functional Study, *Surf. Sci.* **601** (2007), 1747.

14. M. Fuchs, J. L. F. Da Silva, C. Stampfl, J. Neugebauer, and M. Scheffler, Cohesive Properties of Group-III Nitrides: A Comparative Study of All-Electron and Pseudopotential Calculations Using the Generalized Gradient Approximation, *Phys. Rev. B* **65** (2002), 245212.

15. Y. Zhao and D. G. Truhlar, Hybrid Meta Density Functional Theory Methods for Thermochemistry, Thermochemical Kinetics, and Noncovalent Interactions: The MPW1B95 and MPWB1K Models and Comparative Assessments for Hydrogen Bonding and van der Waals Interactions, *J. Phys. Chem. A* **108** (2004), 6908.

16. M. A. Neumann and M.-A. Perrin, Energy Ranking of Molecular Crystals Using Density Functional Theory Calculations and an Empirical van der Waals Correction, *J. Phys. Chem. B* **109** (2005), 15531.

17. G. Pacchioni, Modeling Doped and Defective Oxides in Catalysis with Density Functional Theory Methods: Room for Improvements, *J. Chem. Phys.* **128** (2008), 182505.

18. I. P. R. Moreira, F. Illas, and R. L. Martin, Effect of Fock Exchange on the Electronic Structure and Magnetic Coupling in NiO, *Phys. Rev. B* **65** (2002), 155102.

FURTHER READING

Our discussion of DFT functionals was strongly influenced by J. P. Perdew, A. Ruzsinszky, J. M. Tao, V. N. Staroverov, G. E. Scuseria, and G. I. Csonka, *J. Chem. Phys.* **123** (2005), 062201.

As an entrance point into the recent literature on using hybrid functionals within plane-wave calculations, read J. L. F. Da Silva, M. V. Ganduglia-Pirovano, J. Sauer, V. Bayer, and G. Kresse, *Phys. Rev. B* **75** (2007), 045121, and J. Paier, M. Marsman, K. Hummer, G. Kresse, I. C. Gerber, and J. G. Ángyán, *J. Chem. Phys.* **124** (2006), 154709.

For more information on the derivation of London's force law for nonpolar atoms or molecules, see J. M. Prausnitz, R. N. Lichtenthaler, and E. G. de Azevedo, *Molecular Thermodynamics of Fluid-Phase Equilibria*, 3rd ed., Prentice Hall, Upper Saddle River, NJ, 1999.

DFT-D approaches to including dispersion interactions in molecular calculations based on localized basis sets are described in T. Schwabe and S. Grimme, *Acc. Chem. Res.* **41** (2008), 569, and C. Morgado, M. A. Vincent, I. H. Hillier, and X. Shan, *Phys. Chem. Chem. Phys.* **9** (2007), 448.

DFT+U calculations and several approaches to determining the parameter $(U - J)$ are described in the study by Da Silva et al. cited above and also by N. J. Mosey and E. A. Carter, *Phys. Rev. B* **76** (2007), 155123. Developments and future challenges in this topic area are reviewed in P. Huang and E. A. Carter, *Ann. Rev. Phys. Chem.* **59** (2008), 261.

A comprehensive review of the challenges and methods associated with linear scaling methods is given in S. Goedecker, *Rev. Mod. Phys.* **71** (1999), 1085. For a starting point in learning about ongoing development of codes suitable for linear scaling calculations with bulk materials, see E. Artacho et al., *J. Phys.: Condens. Mat.* **20** (2008), 064208, and C. K. Skylaris, P. D. Haynes, A. A. Mostofi, and M. C. Payne, *J. Phys.: Condens. Mat.* **20** (2008), 064209.

As mentioned in Chapter 9, two excellent sources about classical simulations are M. P. Allen and D. J. Tildesley, *Computer Simulation of Liquids*, Clarendon Press, Oxford, UK, 1987, and D. Frenkel and B. Smit, *Understanding Molecular Simulations: From Algorithms to Applications*, 2nd ed., Academic, San Diego, 2002. For an example of using DFT calculations to parameterize force fields for classical simulations, see A. C. T. van Duin, B. V. Merinov, S. S. Jang, and W. A. Goddard, *J. Phys. Chem. A* **112** (2008), 3133.

INDEX